Dedication

This book is dedicated to my wife, Connie.

For your unwavering love, patience, and belief — through long days, late nights, travel, ideas that would not stop flowing, and even life's unexpected challenges — including a heart attack and bypass that changed perspective forever.

Behind every chapter stands your steady presence. This work exists because of passion — but it endured because of you.

With love and gratitude. Always and forever.

Acknowledgments

No book is ever created in isolation, and this one is no exception. I am deeply grateful to the many individuals who have shaped my thinking, challenged my assumptions, supported my journey, and helped bring this work to life—both in ways seen and unseen.

I would like to begin by recognizing John Petze.
John's influence on this industry—and on my own perspective—has been profound. His vision, leadership, and unwavering commitment to openness and innovation helped shape not only the trajectory of Niagara, but the broader movement that continues to define our space today. On a personal level, his friendship, guidance, conversations, and examples have left a lasting mark on how I think about technology, community, and the responsibility that comes with building something that matters. For that, I am especially grateful.

My sincere thanks as well to the many colleagues, partners, and friends who have contributed along the way: Jerry Frank, Brian Frank, Ed Merwin, Rick Weisensale, Scott Boehm, Scott Muench, James Johnson, Gil Rockwell, John Sublet, Dan Giorgis, Andy Saunders, Gene Allgood, Steve Fey, Julie Hardesty, Jennifer Faulkner-Nance, Greg Barnes, Paul Oswald, Brian Oswald, Scott Cochrane, Paul Czerwin, Geard Huff, Roger Woodard, Terry Casey, Chris Ryan, Jeff Kegley, Gareth Johnson, Ken Smyers, Eric Stromquist, John Donohue, Jack Berryhill, Emily Weisensale, Tom Shircliff, Rob Murchison, Jacob Jackson, Larry Melton, Terry Swope, Melissa Dziurawiec, Pat Hirsch, Robert Hirsch, Tanya Peterson, Charles Johnson, Jim Young, Lisa Woods, and Howard Berger.

Each of you, in your own way, has contributed to the ideas, experiences, and perspective reflected in these pages. I am thankful for your support, your insights, and your role in the journey.

Each of you, in your own way, has contributed insight, encouragement, perspective, and inspiration over the years. Whether through collaboration, mentorship, friendship, debate, or shared purpose, your influence is woven throughout these pages.

I am also profoundly thankful for the **global Niagara Community**—a remarkable network of innovators, integrators, developers, partners, and advocates who have demonstrated what is possible when openness, collaboration, and shared vision come first. This community has not only shaped a framework, but a culture—one built on trust, curiosity, and the belief that we are stronger together. To all of you: thank you for your generosity, passion, and belief in this industry and its people.

Foreword By John Petze

Having had a front row seat in the success and adoption of the Niagara Framework I am excited to offer a foreword for this important book. The genesis and impact of the Niagara Framework journey is a story that needs to be told.

The introduction of computer technology to the control and automation of buildings, which started in the very late 1970's and early 1980's, transformed the field and brought it into the digital technology age.

Throughout the last 5 decades there have been numerous technology advances that serve to define eras of the industry; starting with the initial implementation of building controls on mainframes (70's and early 80's"), to the adoption of micro-controllers and microprocessors (early 80's), to the first generation of primitive, yet functional, networking which enabled distributed control (80's), to the adoption of web based technologies (late 90's – early 2000's). The adoption of digital technology allowed the industry to follow the advance of computer technology, albeit often a generation behind.

What stayed true through all those generations, however, was the closed ecosystem approach of all the major vendors. While their product technology advanced, those products only worked with their own branded equipment. In those early stages this appeared adequate and was accepted.

In the late 90's a small group of people, influenced by backgrounds in advanced communications systems, rapidly advancing web-based software capabilities, including the soon to be ubiquitous web browser, and frustration with the limitations of single-vendor product lines, thought to take a very different approach. By embracing 3 fundamental differences from the status quo; web-based user interfaces, IP-based networking and most importantly the ability to communicate with multiple vendors products, this new approach "normalized" diverse systems providing the user with a uniform, inherently simplified view of their buildings.

To say this was revolutionary is still a bit of an understatement. Those things "just weren't done" at the time. The mission of engineering teams at established controls companies was "to build the next generation proprietary product line". I know, having personally been given those marching orders. As a side note, that radical divergence from the status quo led to the company's first tag line – "revolutionary software'.

Niagara ushered in a bold reinterpretation of the structure, capabilities, and mission of building automation systems and the communities that implemented them. The technology in and of itself was not enough to drive the change though.

Building the ecosystem and reach of the brand required educating owners, operators, consulting engineers and OEMs – the entire value chain – about what was now possible and the benefits this new approach provided.

This book provides perspective and insight into the unique set of circumstances, technologies and branding strategies that led to what is now the world's most adopted software platform in the building automation industry – the Niagara Framework – and it provides context to appreciate the fundamental changes it brought about in the expectations of what building automation should be.

From across the industry...

"Every industry has top tier leaders, and the smart building industry is no different. Marc Petock has delivered consistent vision and leadership over a long period of time and has devoted his life to helping our industry improve. Hopefully, this thoughtful book will serve as a guide and inspiration to our current industry professionals as well as future generations to come."

Jim Young
Founder/CEO Realcomm Conference Group

"Marc Petock has an extraordinary gift for turning complex ideas into clear, actionable insights. His work is not just informative—it's transformative. With a rare blend of vision, expertise, and authenticity, Marc inspires us to think bigger, act smarter, and embrace innovation with confidence. This is a must-read for anyone ready to navigate the future with clarity and purpose."

Terry D Swope
CEO, Chairman, Lynxspring, Inc.

"Marc Petock brings his insight, passion and industry leading vision together at a pivotal time in our industry. I've often referred to Marc as one of the "Godfathers" of the Smart Buildings industry, and his institutional knowledge is critical for the future workforce. Sharing his vision and experience is not only appreciated but a gift! Thank you, Marc, for all that you do for so many of us."

Lawrence Melton
CEO & President, The Building People

"Marc is uniquely qualified to provide this essential historical perspective, given his depth of experience, domain expertise, and longstanding credibility in the industry. He is widely regarded as a leading voice in smart buildings. Those working in this space understand that operational technology differs from traditional IT both technically and culturally. That distinction is increasingly important in the age of AI, making this a valuable read for all stakeholders involved in the built environment."

Tom Shircliff, CRE®
Otterview Partners, LLC

Prologue

There was a time when building automation didn't have a brand. It had cabinets. It had relays. It had proprietary software protected by vendor-specific controllers. It had technicians who guarded knowledge like trade secrets. But it did not have a story.

When I first entered this industry, I didn't walk into a marketplace filled with visionary platforms and strategic positioning statements. I walked into mechanical rooms. Loud ones. Hot ones. Rooms where building systems lived quietly behind walls while the business world above rarely thought about them. If the temperature was comfortable, no one asked questions. If the system fails, someone made a phone call. That was the industry.

We weren't thinking about brand. We were thinking about sequences, control loops, point counts, and commissioning schedules. Brand was something consumer companies cared about — sneakers, soda, or software you downloaded on your home computer.

But over time, I began to see something else happening. Certain names began appearing repeatedly in specifications. Certain platforms began being spoken about differently. Certain companies were trusted before the proposal was even opened.

And that's when I realized something important: Even in mechanical rooms, brand matters.

Preface **Why Brand Matters in an Industry That Never Thought It Did**

For most of my career in building automation and controls, I rarely heard anyone talk about brand that is until I got here.

Engineers talked about specifications. System integrators talked about installation and commissioning. Building owners talked about operational outcomes. Manufacturers talked about products and features.

But brand? That was rarely part of the conversation. In fact, for many years the building automation industry viewed brand as something closer to marketing language than strategic infrastructure. The assumption was simple: if the technology works, the brand will take care of itself.

In many industries, that might be true. But the more time I spent working in the world of building automation, controls, and smart buildings, the more I realized something important. Infrastructure industries depend on brand far more than most people recognize.

Buildings operate for decades. Automation systems often remain in place for twenty years or longer. The technology decisions made today will influence how buildings function for an entire generation. When systems are deployed at that scale and for that length of time, organizations are not simply choosing products.

They are choosing **architectures**. And architecture decisions require something deeper than features or performance specifications.

They require **trust**. Trust that the technology will endure. Trust that the ecosystem supporting it will continue to grow. Trust that the architecture will remain relevant as the industry evolves.

In infrastructure industries, brand becomes the signal that trust exists. Not because of advertising campaigns. Not because of logos or messaging. But because over time, thousands of professionals begin to rely on the same technology, the same architecture, and the same ecosystem.

When that happens, something interesting occurs. Technology stops being evaluated purely on technical merit. It becomes something more foundational. It becomes **infrastructure**.

And infrastructure, when widely trusted, becomes **brand**.

Over the past 30 years, I have had the privilege of being part of and at the center one of the most interesting brand stories in the building automation industry unfold. The story of the Niagara Framework.

When Niagara first appeared, it was not introduced as a brand. It was introduced as a technology framework designed to solve a very real problem: how to integrate disparate building systems that were never designed to communicate with one another.

At the time, buildings were dominated by proprietary architectures. Systems were isolated. Data was trapped. Integration was few and difficult.

Niagara introduced a different idea. A framework that could connect systems rather than replace them. A framework that could enable innovation rather than restrict it. A framework that could allow multiple manufacturers, integrators, engineers, and developers to work within a shared environment.

Over time, something remarkable happened. The framework gained adoption. The ecosystem grew. A community formed around it. And gradually, Niagara became more than a technology. It became a movement. Then a platform. And eventually, a brand.

This book explores how that happened. But more importantly, it explores **what the building automation industry can learn from it.** Because the story of Niagara is not just the story of one technology platform. It is a story about how industries evolve. How ecosystems form. How trust develops. And how infrastructure technologies eventually become foundational to entire markets.

The building automation and controls industry is entering a period of enormous change. Buildings are becoming increasingly connected. Operational technology is converging with enterprise information systems. Artificial intelligence is beginning to influence building operations. Energy systems are becoming more dynamic and decentralized. And the role of data in the built environment is expanding rapidly.

In this new environment, architecture matters more than ever. Platforms matter. Ecosystems matter. And brand is the signal of long-term trust in infrastructure and matters more than most people realize.

My goal in writing this book is not simply to document the history of the Niagara Framework. It is to explore a broader question:

How do technologies become trusted foundations of industries?

What transforms a technical framework into a global ecosystem? What allows a technology platform to endure decades of technological change? And what lessons can the building automation industry learn from that journey?

The answers to these questions extend far beyond any single company or platform. They speak to the future of how buildings will be designed, integrated, and operated. And they speak to the importance of building technology ecosystems that are open, adaptable, and capable of evolving with the needs of the built environment. Because ultimately, buildings are not static systems. They are living infrastructure. And the technologies that support them must be able to evolve alongside them.

The Niagara story offers a remarkable example of how that evolution can happen. It demonstrates the power of openness. The importance of community. And the enduring value of architectural frameworks that enable innovation rather than constrain it.

But perhaps most importantly, it demonstrates something that many industries overlook. The most powerful brands in infrastructure industries are not built through marketing campaigns. They are built through **shared experience, trust, and collective participation.** They are built by the people who design systems. The people who integrate them. The people who operate buildings every day.

This book is written for those people. The engineers. The system integrators. The building owners and operators. The developers. The innovators who continue to push the built environment forward.

Because ultimately, the future of buildings will not be defined by individual products. It will be defined by platforms, ecosystems, and the communities that build them.

And that story is only just beginning.

— **Marc Petock**

Contents

Table of Contents

Foreword By John Petze 6

Preface Why Brand Matters in an Industry That Never Thought It Did 11

Contents 16

Author's Introduction 18

Chapter 1 When Building Automation Was Just Controls 24

Chapter 2 The Evolution of Building Automation 30

Chapter 3 Why Brand Matters in an Industry That Historically Ignored It 38

Chapter 4 Origin and History of the Niagara Framework® 42

Chapter 5 Breaking the One Vendor = One System Mindset 51

Chapter 6 How Niagara Solved the Opportunity Gap 59

Chapter 7 The Niagara Doctrine 67

Chapter 8 The Niagara Movement 72

Chapter 9 The Ecosystem Effect — How Communities Build Great Brands 79

Chapter 10 From Framework to Flagship — How Niagara Became a Brand 86

Chapter 11 Niagara Joined the Great Brands 92

Chapter 12 Why Niagara Has Endurance 99

Chapter 13 Lessons for the Industry — What Can We Learn from Niagara 106

Chapter 14 What Niagara Means for the People Who Actually Build and Run Buildings 114

Chapter 15 What Brand Really Is 120

Chapter 16 The Brand Promise 125

Chapter 17 Brand Loyalty and the Human Layer 131

Chapter 18 Brand as Strategic Capital 136

Chapter 19 What I Learned: How a Framework Becomes a Brand 140

Chapter 20 What Owners Actually Want 146

Chapter 21 When Infrastructure Becomes Identity 155

Appendix The Petock Smart Building Thesis 162

Author's Introduction

"I have always believed in the immense power of brand. Its significance cannot be overstated. A brand shapes perception, builds loyalty, protects value, and influences long-term outcomes. In saturated markets, brand becomes decisive." – **Marc Petock**

Many technologies that shape our world are invisible. We do not think about them every day. We rarely see their logos. And yet they quietly power the systems that modern life depends on. These technologies rarely become famous in the way consumer products do. But they often become something more powerful.

They become infrastructure. Infrastructure technologies have unique characteristics. When they work well, they fade into the background. They simply become part of the environment, trusted, relied upon, and rarely questioned. But when you look closely at these technologies, you notice something interesting.

Many of them have become brands. Not brands in the traditional marketing sense, but brands in the sense that professionals trust them, specify them, build around them, and design entire industries upon them.

In the world of the built environment, one such technology is the **Niagara Framework®**.

The Problem Niagara Solved

To understand why Niagara matters, it helps to look at the building automation and controls industry before it existed.

For decades, buildings were controlled by siloed systems. HVAC systems operated independently. Lighting systems operated independently. Security systems operated independently Energy systems operated independently.

Each system had its own controllers, software, and network architecture. Integration between these systems was often difficult and sometimes impossible. The result was fragmentation.

Data remained trapped inside proprietary environments. Operational insight was limited. And innovation was slowed because every system lived inside its own technological boundaries. This was the state of the industry for many years.

A Different Idea

The Niagara Framework introduced a fundamentally different approach. Instead of trying to replace every building system, Niagara focused on connecting them. It introduced a software framework capable of integrating multiple protocols, devices, and systems into a unified operational environment.

In simple terms, Niagara became a common language for building systems. HVAC systems could communicate with lighting systems. Energy systems could communicate with security systems. Data from multiple systems could be normalized, visualized, and analyzed within a shared environment.

This architectural shift was subtle at first, but its impact was profound. For the first time, building automation could move beyond isolated control systems and toward integrated operational environments. Buildings were no longer just mechanical environments. They were becoming digital environments.

From Framework to Ecosystem

Technologies rarely change industries on their own. What changes industries is what happens around the technology.

Manufacturers began embedding Niagara into their devices. System integrators began using it to connect complex building systems. Consulting engineers began specifying Niagara-based architectures. Developers began building software applications that extended the platform. Building owners began deploying Niagara across entire portfolios.

Gradually, a network of participants formed around the framework. An ecosystem. And ecosystems are powerful. They accelerate innovation. They spread knowledge. They create trust. Over time, the Niagara ecosystem expanded to include hundreds of organizations and thousands of professionals around the world.

What began as a technical framework was evolving into something larger. And eventually, a movement.

When Technology Becomes Brand

In most industries, brand is associated with marketing. In infrastructure industries, brand means something very different. Brand becomes shorthand for reliability. For trust. For long-term architectural confidence.

Engineers specify technologies they trust. System integrators build their businesses around platforms they believe in. Building owners deploy architectures that will remain viable for decades. When thousands of professionals consistently make these choices, technology begins to acquire a reputation. And reputation eventually becomes brand.

That is what happened to Niagara. Over time, Niagara became more than a software framework. It became a trusted architectural layer within the building automation industry. It became something people believed in. And still do today.

The Deeper Story

This book is not simply about the Niagara Framework. It is about the broader ideas that Niagara represents. The importance of open architecture in infrastructure industries. The role of ecosystems in driving innovation.

The relationship between trust and brand. And the transformation of buildings from isolated mechanical systems into connected digital environments. Niagara provides a fascinating lens through which to explore these ideas.

Because its evolution mirrors the evolution of the building automation industry itself. From proprietary systems to open frameworks. From integration challenges to interoperable ecosystems. From control systems to intelligent building platforms.

The Industry at a Turning Point

Today, the built environment is entering another major transition. Buildings are becoming more connected than ever before. Sensors, IoT devices, analytics platforms, and artificial intelligence are rapidly expanding the capabilities of building operations. Energy systems are becoming more dynamic and decentralized. Operational technology is converging with enterprise information systems. Data is becoming the most valuable resource within the building infrastructure.

In this environment, architecture matters more than ever. The technologies that succeed will not simply control equipment. They will provide the framework that allows innovation to occur. They will enable ecosystems. They will unlock data. And they will allow buildings to evolve over time.

The Purpose of This Book

This book explores how one framework helped reshape the architecture of building automation and controls. But more importantly, it explores the lessons that journey offers. Lessons about openness. Lessons about ecosystems. Lessons about how infrastructure technologies become trusted foundations of industries.

It also explores a broader question: ***What does the future of building automation and controls look like when buildings become fully connected, data-driven environments?***

And what role will platform architectures play in enabling that future? Because the next phase of the built environment is already beginning. Buildings are moving beyond control and integration toward intelligence and eventually autonomy.

The technologies that guide that transition will not just be products. They will be platforms, ecosystems, and trusted frameworks.

And understanding how those platforms emerge and why professionals trust them may be one of the most important conversations the building automation and controls industry can have.

A Journey Through Technology, Ecosystems, and Brand

The chapters that follow explore this journey. How Niagara introduced a new approach. How ecosystems form around an open framework. How infrastructure technology became brand.

The story of Niagara is still being written. But the ideas introduced openness, integration, community, and platform thinking – will likely shape the built environment for decades to come.

And that makes it one of the most interesting stories in the modern history of building automation and controls.

Chapter 1 **When Building Automation Was Just Controls**

There was a time when building automation was not really an industry. It was a craft.

It lived in mechanical rooms and electrical closets. It spoke the language of relays, pneumatics, voltages, and proprietary protocols. It was practiced by specialists who carried vendor-specific software on their laptops—often protected by hardware dongles and guarded passwords.

Systems were installed, commissioned, and, if everything went well, they quietly disappeared into the background. Buildings were automated, but they were not connected. They were controlled, but they were not integrated. They certainly were not intelligent.

And they were not branded.

Before the emergence of modern platforms, most building automation systems existed as vertically integrated silos: one manufacturer, one tool, one ecosystem, one way of thinking. If you installed Vendor A, you used Vendor A's workstation. If you switched to Vendor B, you started over.

If your integrator changed, your system became fragile. Knowledge did not transfer. Experience did not scale. Loyalty was not earned through value—it was enforced through technology and closely held knowledge.

At the time, this seemed normal. It was simply the way the industry worked.

The Era of Proprietary Control

Early building automation vendors built complete solutions: controllers, software, graphics, networking, and services. Everything was designed to work together inside the vendor's ecosystem.

Internally, the systems were optimized. Externally, they were not.

Protocols were inconsistent. Tools were incompatible. Documentation was fragmented.

Training was vendor-specific. Certification applied only to a single product line. An integrator supporting five manufacturers might carry five laptops, five software licenses, and five entirely different mental models of how systems worked.

The industry accepted this complexity—not because it was efficient, but because there was no alternative.

Buildings as Isolated Machines

In those years, buildings were treated much like mechanical machines. Once installed, they were expected to run.

The philosophy was simple: set it and forget it.

If something failed, a technician was dispatched. Optimization was manual. Data remained local. Reporting was primitive.

Remote access was rare and often avoided. Systems were designed to control equipment—not to produce insight. The model provided stability, but not adaptability. It worked well enough for a world where buildings were largely static and expectations changed slowly.

Because buildings were managed this way, brand carried little meaning.

No one asked, "What platform are you on?" Instead, they asked a much simpler question: "Does it work?"

If the answer was yes, the conversation ended there.

An Engineering Culture, Not a Brand Culture

The building controls industry was built by engineers.

Engineers solve problems. They design systems. They write sequences. They troubleshoot failures. But they rarely tell stories. They do not think first in terms of perception, identity, or emotional connection. They think in terms of reliability, specifications, schematics, and time between failures.

This shaped the culture of the industry.

Technical mastery mattered more than market identity. Reputation existed—but brand did not.

Companies were known because their systems worked, their technicians showed up, and their drawings were accurate. Word of mouth carried credibility far more effectively than marketing ever could.

There were no positioning statements. No brand playbooks. No narratives about transformation or strategy.

There were simply good contractors and unreliable ones. Reliable manufacturers and problematic ones.

That was the entire reputation economy of the industry.

The Dominance of Proprietary Thinking

Perhaps the most defining feature of this era was the proprietary mindset.

Most systems were intentionally closed. Protocols were guarded. Software access was restricted. Tools were limited to certified technicians. Data was rarely discussed. Cybersecurity was not yet part of the conversation.

In many ways, the industry created small monopolies inside buildings.

Once a system was installed, the owner was effectively locked in. Upgrades, expansions, and service flowed through a single provider. From a business standpoint, this model worked well.

From a brand standpoint, it made brand unnecessary.

Why invest in brand when customers had nowhere else to go? Why build loyalty when switching was prohibitively expensive? Why tell a story when the contract—and the technology—did the talking?

Selling building automation in this era was fundamentally transactional. Proposals and RFI/RFPs focused on point counts, equipment lists, installation schedules, and maintenance contracts.

And, of course, the four-letter word that dominated every conversation: cost.

The Invisible End User

In the controls-only era, the end user was largely invisible.

Building occupants rarely interacted directly with automation systems. Facility managers acted as intermediaries. Integrators were gatekeepers. Manufacturers were distant suppliers.

The user experience was not designed. It was tolerated.

If a building was too hot or too cold, someone called maintenance. If a system failed, a technician was dispatched. If energy was wasted, it was accepted as part of doing business.

Systems lived behind walls, above ceilings, and inside locked closets—out of sight and out of mind.

The First Cracks in the Wall

Eventually, technology began to shift.

Microprocessors replaced relays. Software replaced hard logic. Networks replaced point-to-point wiring. Buildings began generating data. Operators started seeing trends. Owners began asking questions about energy use and operational performance.

At the same time, customers became more comfortable with technology, and corporate IT departments began paying attention. Campuses wanted centralized monitoring. Enterprises wanted portfolio-wide visibility. Integrators wanted fewer tools. Users wanted an experience that matched the software they used elsewhere.

The old proprietary systems struggled to support these demands.

They had been designed to protect territory—not enable ecosystems.

The walls of the siloed model began to crack.

When Control Became Information

As networking expanded, automation stopped being just about control. It became about information.

And when data appears, value shifts.

When value shifts, perception begins to matter.

And when perception matters, brand eventually becomes important.

This transformation did not happen overnight. The old mindset lingered—and in some places, it still does.

Even today, some organizations believe that if their product is good enough, brand does not matter. For a long time, that belief was true. In a world of isolated buildings, limited data, and captive customers, brand offered little advantage.

But that world no longer exists.

Technology accelerated. Markets expanded. Customers demanded choice—and a more modern experience. Data became power. Systems began talking to each other.

And once buildings became data engines, control alone was no longer enough.

The industry was about to change.

And with it, the role of brand would change forever.

Chapter 2 **The Evolution of Building Automation**

The story of building automation is, at its core, a story of technological maturation.

What began as simple mechanical control has evolved into something far more consequential—a digital infrastructure that increasingly resembles an operating system for the built environment.

Buildings today are expected to do far more than regulate temperature and keep equipment running. They are expected to manage energy, optimize performance, protect cybersecurity, support sustainability goals, enhance occupant experience, and provide enterprise-level insight across entire portfolios.

That's a different mandate entirely.

This transformation did not happen overnight. It unfolded in distinct phases, each driven by technological advancement, shifting industry needs, and rising expectations.

Understanding this evolution is not just about history. It explains why the industry behaves the way it does today—and why it is still catching up to what buildings are now expected to deliver.

Phase 1 — Mechanical Control Era

"Buildings That React"
1970s–1990s

The earliest era of building automation focused on one objective: control.

Systems were designed to regulate temperature, operate HVAC equipment, and maintain basic environmental stability. Compared to pneumatic systems and manual operation, this was a major step forward.

But the capabilities—and the mindset—were limited.

These systems were built as stand-alone solutions. HVAC, lighting, security, and energy systems operated independently, each within its own closed environment. Integration was not expected, and in many cases, not even possible.

The industry leaned heavily toward mechanical engineering principles. Systems were local. Data stayed inside the panel. And once installed, buildings were expected to run.

Vendor lock-in wasn't a side effect. It was the model.

Choosing a vendor meant committing to their tools, their software, and their way of thinking. Switching providers often meant replacing large portions of the system.

In many ways, buildings during this era were automated—but not connected.

They reacted to conditions, but they did not learn from them. Data existed, but it had no real pathway to influence broader operations. Intelligence, as we think of it today, simply wasn't part of the equation.

Phase 2 — Digital Integration Era

"Buildings That Connect"
2000s–2010s

The next major shift came when networking technologies began reshaping building infrastructure.

As portfolios expanded and expectations grew, the limitations of isolated systems became impossible to ignore. Owners needed visibility across systems. Integrators needed flexibility. And the industry needed a way to connect technologies that had never been designed to work together.

This was the beginning of true integration.

Open protocols such as BACnet, Modbus, and LonWorks made cross-system communication possible. Web-based interfaces changed how systems were accessed and managed.

But protocols alone didn't solve the problem.

They allowed communication—but not coordination.

That's where integration frameworks changed the game.

Platforms like the Niagara Framework introduced a standardized layer capable of connecting diverse systems into a unified environment. For the first time, integrators could design architectures that spanned multiple vendors without being constrained by any single one.

This was more than a technical shift. It was a structural one.

Buildings were no longer collections of independent systems. They became connected environments capable of sharing data and supporting coordinated operation.

And with that, the industry began moving—sometimes reluctantly—away from proprietary isolation toward platform-based thinking.

Some embraced it. Others resisted it.

But the direction was set.

Phase 3 — Intelligent Building Era

“Buildings That Learn”
2020s

Today, the industry has entered a new phase—one defined not by control or connectivity, but by intelligence.

The rapid growth of cloud computing, edge processing, and advanced analytics has shifted the focus of building automation toward data-driven performance.

Modern buildings generate enormous volumes of operational data—from HVAC systems and energy meters to occupancy sensors and equipment diagnostics. The challenge is no longer connecting systems.

It is extracting value from the data those systems produce.

Technologies like edge computing, cloud analytics platforms, and artificial intelligence are enabling a different operational model—one that moves beyond reactive responses toward predictive and outcome-driven strategies.

Maintenance can be anticipated before failures occur. Energy usage can be optimized dynamically. Portfolio-wide insights can reveal inefficiencies that would have gone unnoticed in isolated systems.

In this phase, buildings begin to function less like mechanical systems and more like data platforms.

Operational technology is no longer separate from enterprise IT—it is becoming part of the same conversation. Building data now informs decisions about energy strategy, sustainability initiatives, asset planning, and financial performance.

The building is no longer just infrastructure.

It is becoming part of a broader digital ecosystem.

Phase 4 — Autonomous Building Era

"Buildings That Adapt"
Emerging

The next phase is already beginning to take shape.

Autonomous buildings move beyond analysis and optimization toward self-directed operation. Systems don't just provide insight—they act on it.

Artificial intelligence, predictive modeling, and advanced edge analytics are enabling buildings to adapt dynamically to changing conditions without constant human intervention.

Digital twins will simulate performance and test strategies before they are deployed in real time. AI systems will adjust HVAC operation based on weather patterns, occupancy trends, and energy pricing. Maintenance will shift from scheduled intervals to condition-based execution.

Buildings will no longer simply react to conditions—or even learn from them.

They will adapt.

That does not eliminate the role of human operators. It changes it.

Operators move from reactive maintenance to strategic oversight. The focus shifts from fixing problems to managing systems that are continuously optimizing themselves.

This is where automation begins to cross into autonomy.

And it will not happen evenly. Some buildings will move quickly. Others will lag years behind.

But the shift is underway.

Strategic Insight: A Layered Evolution

This evolution is not a replacement cycle. Each phase builds on the one before it.

Control enabled automation. Integration connected systems. Intelligence turned data into insight. And autonomy is beginning to turn insight into action.

That progression can be understood simply:

Control → Integration → Intelligence → Autonomy

And here's the reality most people underestimate:

All four phases exist at the same time.

Some buildings are still largely mechanical. Others are integrated but not intelligent. A smaller group is beginning to operate as truly data-driven environments.

That fragmentation explains much of the confusion in the industry.

Different stakeholders are solving different problems—based on the phase they're operating in.

Within this progression, certain technologies have played outsized roles.

Integration frameworks—most notably the Niagara Framework—helped define the integration era by enabling multi-system connectivity and open ecosystem development. Today, those same platforms continue to serve as foundational infrastructure for data collection, analytics, and enterprise integration.

They didn't just solve a problem.

They changed the direction of the industry.

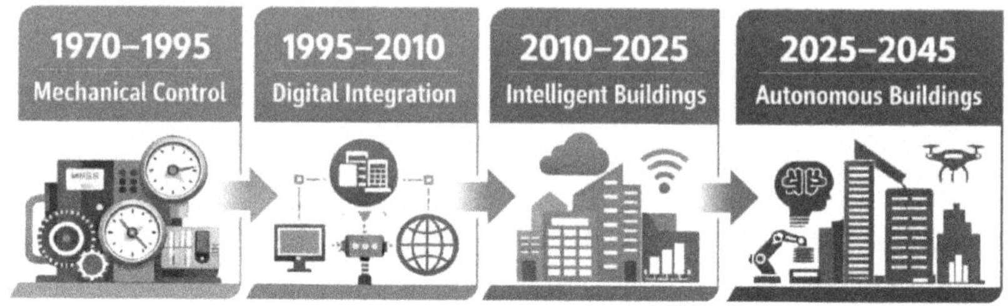

Figure 1: Evolution of Building Automation and Controls

> *The building automation industry has progressed from isolated mechanical control to integrated, data-driven, and increasingly autonomous systems. Each phase builds on the last, but all still coexist in practice.*

What This Really Means

The journey from mechanical control to intelligent infrastructure is one of the most significant transformations in the history of the built environment.

And it's not finished.

The next generation of buildings will not simply be automated or integrated. They will be adaptive systems—capable of learning, optimizing, and responding dynamically to the needs of occupants, energy systems, and an increasingly connected world.

That shift raises a different kind of question.

Not “what can the system control?”

But:

“What can the building understand—and what can it do with that understanding?”

The answer to that question will define the next era of the industry.

Chapter 3 **Why Brand Matters in an Industry That Historically Ignored It**

For decades, the building automation industry operated under a simple assumption: if the system worked, nothing else really mattered.

Performance, reliability, and price dominated purchasing decisions. Integrators selected the systems they knew. Engineers specified what was familiar. Owners trusted what had already been installed. In that environment, brand rarely entered the conversation—not because it was dismissed, but because it was unnecessary.

The structure of the market reinforced that reality. Decisions were made locally, often at the project level, by contractors, consulting engineers, and facility managers. Loyalty was personal, built on relationships and experience rather than institutional strategy. Systems were largely invisible to the broader organization. When they worked, they disappeared into the background. When they failed, someone made a phone call.

Longevity amplified this dynamic. Building systems remained in place for decades, discouraging experimentation and rewarding conservative choices. The safest decision was the one that had already proven itself. In that context, competence mattered. Identity did not.

That model worked for a long time.

It no longer does.

The shift began as buildings themselves began to change. What had once been isolated mechanical environments started to evolve into connected systems. Integration increased. Data became more accessible. Expectations expanded beyond basic control toward performance, efficiency, and visibility across entire portfolios.

Building automation stopped being just about equipment. It became part of a broader digital infrastructure—one that connects physical systems with enterprise operations, analytics platforms, and business decision-making.

That change raised the stakes.

Modern building systems now influence energy strategy, sustainability outcomes, cybersecurity posture, tenant experience, and financial performance. In many organizations, they directly affect operational resilience and long-term asset value. Decisions that were once technical are now strategic.

At the same time, the technology itself matured. Most major platforms can now integrate systems, provide data visibility, and connect to enterprise environments. Capabilities that once differentiated vendors have begun to converge.

When that happens, the basis of competition changes.

Buyers are no longer asking, "Can this system do the job?" They are asking, "Who do I trust to support this over the next ten to twenty years?"

That question reframes everything.

It introduces concerns that did not exist in the earlier model. Will the platform remain supported over time? Will cybersecurity evolve alongside emerging threats? Will integrations survive the next wave of technological change? Will the company remain committed to openness, or shift direction once it has market share?

These are not feature comparisons. They are assessments of risk.

And risk is where brand enters the equation.

In the modern built environment, brand functions as a signal of future reliability. It represents confidence in how a platform will perform not just today, but over its entire lifecycle. It reflects whether an organization has demonstrated the discipline, consistency, and alignment required to support long-term infrastructure decisions.

This is fundamentally different from how brand is often understood. It is not about recognition or messaging. It is about reducing uncertainty in environments where the cost of being wrong is high.

When systems were isolated and expectations were limited, that uncertainty was manageable. Switching costs were high, but the scope of impact was relatively contained. Today, the consequences of a poor decision extend far beyond a single system. They affect data strategies, integration architectures, cybersecurity exposure, and long-term operational flexibility.

As risk increases, decision-makers look for signals they can trust.

Brand becomes one of the most important signals available.

Not because it is promoted, but because it is observed.

Parts of the industry are still adjusting to this reality. The legacy mindset—competing on features, price, and familiarity—persists in many areas. That approach was effective in a slower, more isolated market. It is less effective in a platform-driven one.

Because when buyers are making long-term infrastructure decisions, they are not just selecting a system. They are aligning with a roadmap, a philosophy, and a pattern of behavior that will shape how their buildings evolve over time.

Whether organizations acknowledge it or not, those decisions are being made through the lens of brand.

Looking forward, this dynamic will only intensify. Buildings will become more connected, more data-driven, and increasingly integrated with enterprise systems. The expectations placed on building infrastructure will continue to expand.

Technology will continue to advance. Capabilities will continue to converge.

And differentiation will increasingly come from something less visible, but more durable: the ability to deliver consistent outcomes over time.

That is what brand represents in this industry.

And in the next phase of building automation, it will determine which platforms are trusted to move forward—and which are left behind.

Chapter 4 **Origin and History of the Niagara Framework®**

The Idea That Changed Everything

In the early 1990s, building automation was not shaped by platforms, data, or interoperability. It was a collection of independent control systems installed to manage individual mechanical and electrical functions inside buildings.

HVAC operated on one system. Lighting on another. Metering, security, and fire protection each existed in separate technological domains.

These systems rarely communicated. When integration was attempted, it required custom gateways, proprietary tools, and painstaking manual programming—and even then, the results were fragile.

The problem was not the hardware.

It was the architecture.

Each manufacturer had built a complete, vertically integrated ecosystem—controllers, software, networking, tools, and services. These systems worked well within their own boundaries, but struggled outside them. Buildings were controlled, but they were not connected.

As facilities grew more complex, this fragmentation became increasingly difficult to manage. Data could not move freely. Visualization was limited and inconsistent. Scaling across portfolios became expensive and time-consuming.

The industry had become very good at installing controls.

It was inefficient at managing information.

A Radical Idea

During this period, a small team at Tridium in Richmond, Virginia began looking at the problem differently.

They asked a simple—but radical—question:

What if buildings operated on a common software framework, independent of vendor, protocol, or hardware?

The inspiration came from computing. Personal computers became exponentially more useful once operating systems created a shared environment where applications and devices could interact.

What if buildings worked the same way?

Instead of building another proprietary control system, Tridium set out to build something fundamentally different: a software infrastructure layer capable of connecting disparate devices, normalizing their data, and presenting them through a unified interface.

This was not another product.

It was a framework.

That distinction would define everything that followed.

From Prototype to Platform (1997–1999)

By 1997, the team had developed a working prototype capable of connecting devices from multiple manufacturers, translating between protocols, and presenting data through a web-based interface.

At the time, this was a sharp departure from industry norms. Most systems required proprietary workstations and specialized software. The idea that building systems could be accessed through a standard web browser was, at the time, disruptive.

In 1999, Tridium formally launched the **Niagara Framework®** with its first commercial release: Niagara R2.

From the beginning, the framework was built on a set of principles that were uncommon in the industry:

It was protocol agnostic, vendor neutral, and network native. It embraced object-oriented architecture and was designed to be extensible.

More importantly, it focused on the integration layer—the point where systems meet, data converges, and information becomes usable.

That architectural decision changed the trajectory of the industry.

Early Adoption and Strategic Alignment (2000–2004)

The early 2000s validated the concept quickly.

In 2000, Tridium entered a strategic alliance with a major OEM manufacturer, giving Niagara immediate global reach at a critical moment. Around the same time, the **Vykon®** brand was introduced as a commercial implementation of the framework—demonstrating that Niagara could be deployed at scale without being tied to a single hardware ecosystem.

Integrators recognized the shift almost immediately.

With Niagara, they could build integration logic once and deploy it across multiple environments. Their expertise became portable. Their intellectual property no longer lived inside a single vendor's ecosystem.

For the first time, the integrator—not the manufacturer—held the architectural center of the project.

That was a fundamental shift in power.

In 2004, Tridium hosted the first Niagara Summit. What began as a user conference quickly evolved into something more significant—a gathering point for a growing ecosystem.

Developers, integrators, OEMs, and partners came together not just to learn, but to align.

Niagara was no longer just a framework.

It was becoming a community.

Reinvention: The Niagara AX Era (2005–2014)

By the early 2000s, Niagara R2 had proven the concept. But platforms that don't evolve don't survive.

In 2005, Tridium introduced Niagara AX—a complete architectural redesign that marked the second generation of the platform.

This was not an incremental upgrade. It was a reinvention.

Niagara AX introduced a more robust component architecture, improved development tools, expanded APIs, stronger enterprise scalability, and enhanced security models.

During this period, adoption accelerated rapidly.

New generations of **JACE (Java Application Control Engine)** controllers expanded deployment options. OEM partnerships multiplied. International adoption grew.

Just as important, a third-party ecosystem began to take shape—communication drivers, applications, and tools developed by partners rather than a single vendor.

By the late 2000s, Niagara had become embedded in mission-critical environments around the world: hospitals, universities, airports, data centers, and large commercial portfolios.

If systems needed to talk to each other, Niagara was often the answer.

Institutionalization and Scale

In 2005, Honeywell acquired Tridium.

That moment could have changed Niagara's trajectory.

It didn't.

Tridium remained operationally independent, and Niagara continued to follow its open framework philosophy. It remained multi-OEM, protocol neutral, and community-driven.

As a result, adoption accelerated.

By the late 2000s and early 2010s, Niagara crossed an important threshold. It was no longer an emerging technology.

It was becoming an industry standard.

The Great Modernization: Niagara 4 (2015–Present)

By the early 2010s, the technology landscape had shifted again.

Cloud computing was reshaping IT. Cybersecurity had become critical. Mobile devices were everywhere. Web technologies had matured. IT departments were now deeply involved in operational environments.

Niagara had to evolve.

In 2015, Tridium introduced Niagara 4—the most significant modernization in the platform's history.

It introduced modern security models, certificate-based authentication, HTML5 user interfaces, improved development workflows, and enterprise-level management capabilities.

More importantly, it aligned Niagara with modern IT expectations.

It repositioned the platform not just as a building automation tool—but as enterprise infrastructure.

Over time, Niagara 4 expanded into edge computing, lifecycle management, and cloud integration. By 2021, it had surpassed one million active instances worldwide—a level of adoption rarely achieved in industrial software.

Renewal and Discipline

In 2020, the final update of Niagara AX was released. In 2021, AX reached end-of-life.

The transition was difficult.

Thousands of systems required upgrades. Integrators had to retrain. Owners had to modernize infrastructure.

But it was necessary.

Platforms that cling to legacy architecture eventually collapse under their own weight.

Niagara chose renewal over stagnation.

From Platform to Digital Infrastructure

By the early 2020s, Niagara's role had expanded well beyond building automation.

It had become a data broker, a cybersecurity boundary, an edge computing environment, and a cloud integration layer.

It now connects building systems to analytics platforms, digital twins, independent data layers, and artificial intelligence applications.

Buildings themselves have changed as well.

They are no longer isolated assets.

They are nodes in larger digital ecosystems.

And in many ways, Niagara has become the operating system of the built environment.

The Next Evolution: Niagara 5

At the time of this writing, Tridium has announced the next generation: Niagara 5.

It is expected to expand capabilities around cloud-native workflows, API-driven integration, advanced cybersecurity models, DevOps-oriented development, and edge-to-cloud orchestration.

If earlier generations focused on integration and enterprise alignment, Niagara 5 represents something larger:

A transition toward true digital infrastructure for buildings.

Why Niagara Has Endured

Many platforms have entered this market.

Few have lasted.

Fewer still have shaped the industry.

Niagara endured because of a set of principles embedded in its design:

It remained neutral—never tied to a single hardware ecosystem.
It stayed extensible—allowing developers and partners to build on top of it.
It fostered community—through summits, partnerships, and shared development.
It embraced reinvention—each generation evolving deliberately, not incrementally.
And it focused on outcomes—what buildings could become, not just how they were controlled.

From Framework to Institution

Today, Niagara is taught in universities and trade schools. It appears in engineering specifications, procurement policies, and integration standards. It is embedded in millions of square feet of buildings worldwide.

What began as an idea has become infrastructure.

And that's the real story.

Niagara is not just software.

It is a philosophy—the belief that open systems scale, platforms outlast products, and community creates resilience.

Niagara didn't succeed because it was simply better software.

It succeeded because it was designed to become a foundation.

And foundations—when built well—support generations.

Why Registered Marks Such as Niagara Framework® Matter

Brands are built not only through innovation and performance—but through consistency, protection, and credibility. The proper use of registered marks such as *Niagara Framework®* signals to the market that these names represent established, protected, and recognized technologies.

Consistency reinforces trust. It tells customers, partners, consultants, and end users that they are engaging with authentic platforms backed by a defined ecosystem, standards, and long-term stewardship. In building automation—where reliability, interoperability, and lifecycle confidence matter—those signals carry weight.

Used correctly, registered marks help prevent confusion, strengthen brand equity, and preserve reputations built over decades. They reinforce that *Niagara Framework®* is more than software—it is a trusted platform with expectations attached to it.

In short: **consistency creates recognition, recognition builds trust,** and **trust drives adoption.**

® Registered

© Copyright

™ Trademark

Editorial Note on Proper Usage in Print

When referencing registered trademarks in professional publications:

- Use the ® symbol on the first or most prominent mention of the term (e.g., *Niagara Framework®*).
- Subsequent mentions may omit the symbol for readability unless emphasis is required.
- Do not overuse the symbol—clarity matters more than repetition.
- Maintain consistent spelling and capitalization as defined by the trademark owner.
- Include a trademark attribution statement in the front matter or copyright page (e.g., "*Niagara Framework®* is a registered trademark of Tridium, Inc.").
- Avoid using trademarks as generic terms; they should identify a specific product or platform.

These practices balance legal correctness with readability—protecting the brand without disrupting the reader experience.

Chapter 5 **Breaking the One Vendor = One System Mindset**

For decades, building automation followed a simple and rigid formula:

One building. One vendor. One system.

Choosing a provider meant choosing everything that came with it—controllers, front-end software, engineering tools, network architecture, and service model. The ecosystem was inseparable from the equipment.

If you chose Honeywell, you entered Honeywell's world. If you chose Siemens or Schneider Electric, the same rules applied.

The industry was not intentionally restrictive.

It was simply built that way.

Hardware platforms were proprietary. Protocols were closed or inconsistently implemented. Engineering tools were tightly controlled. Integration, when it existed at all, relied on gateways and custom programming. Changing vendors often meant starting over.

Over time, the industry normalized a belief that would shape it for decades:

Choice creates risk. Stability requires loyalty.

The Structural Problem

The "one vendor = one system" model created constraints that were rarely discussed openly—but were deeply felt.

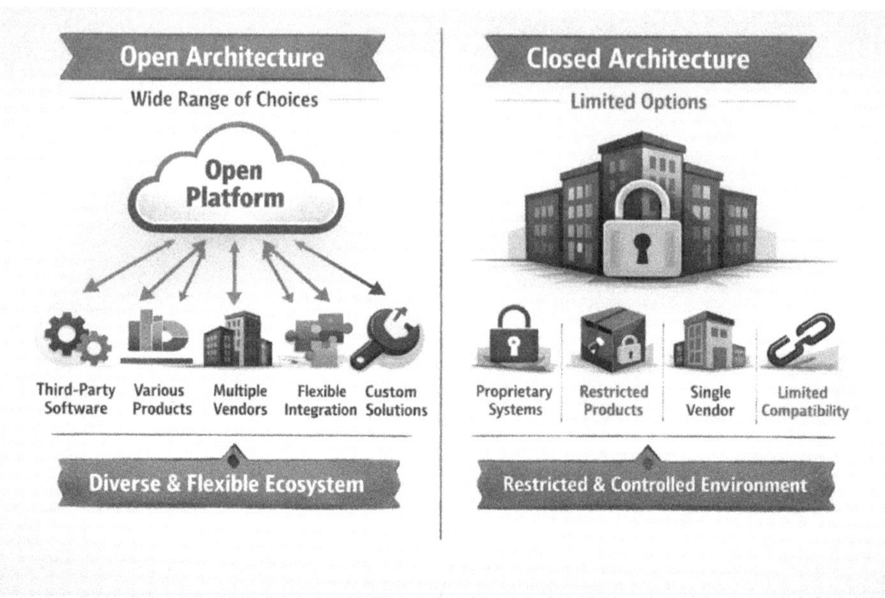

Figure 2: Open vs. Closed Architecture in Building Automation

> *Open architecture enables interoperability, flexibility, and long-term adaptability across systems. Closed architecture restricts integration and reinforces vendor dependency.*

Innovation was tied to a single company's roadmap. If that roadmap stalled, customers had few options.

Pricing power concentrated with the vendor. Once a system was installed, switching costs were so high that competition effectively disappeared.

Expansion reinforced the cycle. New buildings, renovations, and upgrades almost always meant adding more of the same vendor's technology.

Even with the introduction of open protocols like BACnet®, the reality was messier than the theory. Implementations varied. Optional services were interpreted differently. Interoperability existed—but often only at the surface level.

The result was predictable.

Buildings became isolated technology islands. Campuses became patchworks of incompatible systems. Owners became dependent on a single provider for decades.

As buildings grew more complex and more connected, the limitations of this model became impossible to ignore.

What the industry needed was not another controller.

It needed a different architecture.

Enter Niagara

When the Niagara Framework was introduced, it approached the problem from a different angle.

Niagara was not another vertically integrated control system.

It was an integration framework.

Rather than replacing systems, it connected them—bringing together technologies such as BACnet®, Modbus®, LONWORKS®, proprietary protocols, and web services into a common architecture.

At the time, that approach was quietly radical.

Instead of forcing a single-vendor decision, Niagara allowed multiple systems to coexist within a unified operational environment.

That created several structural shifts.

Niagara sat above devices rather than replacing them, allowing legacy systems to remain in place. It normalized data into a common framework, making disparate systems visible and manageable together. And it introduced a shared software environment that multiple manufacturers could adopt.

For the first time, building owners could unify systems across vendors—without committing to one ecosystem.

This wasn't just technical progress.

It was a new way of thinking.

A New Ecosystem Emerges

As adoption grew, a new model began to take shape.

Instead of competing through closed systems, manufacturers began embedding Niagara within their own products. The industry moved—gradually—toward something it had never seen before:

Multiple OEMs operating on a shared framework.

At the center of this shift was Niagara's engineering environment, Workbench.

Before this, engineers had to learn a completely different toolset for each manufacturer. Each OEM had its own workflows, its own logic structure, its own way of doing things.

Workbench changed that.

Once integrators understood Niagara, they could apply those skills across multiple manufacturers. Engineering expertise became portable.

That changed careers.

It changed training.

And it changed leverage.

Integrators gained flexibility. Organizations could standardize engineering practices. Training investments became more durable. And most importantly, the barrier to switching began to drop.

And when switching becomes easier, control shifts.

The Psychological Disruption

The deeper disruption wasn't technical.

It was psychological.

For decades, the industry believed stability required commitment to a single vendor. That belief shaped specifications, procurement strategies, and long-term planning.

Niagara challenged it.

It demonstrated that interoperability and stability were not mutually exclusive. That multiple manufacturers could participate in a shared ecosystem without sacrificing differentiation. That buildings did not have to be technological islands.

Instead, they could become platforms.

As more participants joined the ecosystem, its value increased. Not because of a single company—but because of the network itself.

This is the classic network effect.

And once it took hold, the old model became harder to defend.

Empowering Rather Than Replacing

One of the reasons Niagara succeeded is that it didn't try to force the industry to start over.

It didn't require ripping out existing systems. It didn't attempt to displace every vendor. It didn't impose a single vision.

It created a layer above the existing structure—and made that structure more valuable.

Manufacturers could still differentiate through hardware, applications, and services. Integrators could build multi-vendor solutions. Owners could extend legacy systems while modernizing infrastructure.

Everyone gained something.

And that matters.

Because most disruption fails when it tries to replace everything at once.

Niagara succeeded because it **aligned incentives instead of fighting them**.

Resistance and Momentum

Not everyone welcomed the shift.

Some manufacturers saw shared platforms as a threat. Others worried about commoditization or losing control over customer relationships. There were legitimate concerns about governance and long-term stewardship.

But momentum built anyway.

Integrators saw the efficiency of a unified engineering environment. Consultants recognized the value of specifying open architectures. Owners began demanding flexibility and interoperability.

Over time, market demand—and the clarity of the architecture—won out.

Niagara didn't just grow.

It became expected.

A New Philosophy for the Industry

Niagara ultimately disrupted more than technology.

It disrupted the assumptions behind the market.

For decades, the industry believed vendor loyalty was required for stability.

Niagara demonstrated the opposite:

That openness can create resilience.
That interoperability can reduce risk.
That flexibility can strengthen long-term value.

It enabled multi-OEM participation. It made integrator skills portable. It unified engineering environments. It allowed legacy systems to be modernized without replacement.

But more importantly, it changed how the industry thinks.

Niagara didn't compete by locking customers in.

It competed by giving them options—and the confidence to use them.

And from that confidence, something larger emerged.

Not just a platform.

A brand.

One built on openness, ecosystem thinking, and the belief that buildings should never be limited by the boundaries of a single vendor's system.

Chapter 6 **How Niagara Solved the Opportunity Gap**

For much of its history, the building automation and controls industry lived with a quiet contradiction.

Buildings were becoming more complex. They contained increasingly sophisticated mechanical systems, digital controls, sensors, and software. Yet despite that growing sophistication, most buildings were still operated in fragmented ways.

Systems ran independently. Data lived in silos. Integration was difficult and expensive. Operational insight was limited.

The industry had the tools to automate buildings.

It lacked the architecture to truly connect them.

That disconnect created what can best be described as the **Opportunity Gap** in the built environment. It was not a shortage of equipment. It was not a shortage of data. It was the gap between what buildings were capable of delivering and what operators could realistically achieve.

Over time, one platform fundamentally changed that reality: the Niagara Framework.

Understanding the Opportunity Gap

Modern buildings generate enormous amounts of operational data. Every controller, sensor, meter, and device produces information about temperature, airflow, occupancy, equipment status, alarms, energy consumption, and performance.

In theory, that data should give building owners and operators deep visibility into how their facilities actually function.

But historically, the data remained trapped.

HVAC systems had their own control platforms. Lighting systems were managed separately. Energy meters used different software. Security and access control systems were often isolated entirely. Even when integration was possible, it was expensive, custom-built, and fragile.

The result was predictable: limited cross-system visibility, slow troubleshooting, poor energy optimization, heavy reliance on vendor-specific tools, and little ability to scale operational intelligence across portfolios.

Buildings were automated, but they were not integrated.

And because they were not integrated, much of their potential remained unrealized.

The Architectural Shift

Niagara introduced a fundamentally different approach.

Instead of creating another proprietary control system, Niagara was designed as an integration framework—a platform capable of connecting diverse devices, protocols, and systems into a unified environment.

At its core, Niagara addressed the problem that had limited the industry for decades: interoperability.

It allowed different building technologies to communicate regardless of manufacturer or protocol. It created a common environment where these systems could coexist, exchange data, and be managed together.

That changed everything.

Integration was no longer just a one-off engineering effort performed under stress at the end of a project.

It became a platform capability.

From Silos to Systems

Before Niagara, most buildings were constructed around vertically integrated systems. Each vendor provided its own controllers, software, tools, and services. That approach simplified procurement in the short term, but it created long-term rigidity.

Once installed, buildings became dependent on a single technology ecosystem. Adding new systems, integrating other platforms, or changing service providers became difficult and costly.

Niagara introduced a different model.

Instead of isolated automation systems, buildings could operate as systems of systems. HVAC could coordinate with lighting. Energy systems could inform operational strategies. Security systems could interact with occupancy data. Enterprise software could access building information in real time.

For the first time, buildings could be managed holistically rather than system by system.

That was not just a technical improvement.

It was an operational breakthrough.

Unlocking Building Data

One of Niagara's most important contributions has been making building data usable.

Buildings had always generated data. The challenge was organizing that data in ways that made it accessible, consistent, and meaningful.

Niagara provided the tools to collect, normalize, and structure information across the building environment. Once that data was unified, entirely new capabilities became practical: energy analytics, performance benchmarking, portfolio visibility, predictive maintenance, sustainability reporting, digital twin development, and operational optimization.

What had once been invisible became actionable.

Facilities teams could see patterns across systems. Energy managers could compare performance across buildings. Maintenance teams could identify problems before they became failures.

The industry's focus began shifting from automation as control to automation as intelligence.

Bridging OT and Enterprise Systems

Another critical part of Niagara's impact was its ability to bridge operational technology and enterprise information systems.

Historically, building systems lived outside the enterprise IT ecosystem. They were managed by facilities teams using specialized tools that rarely interacted with business systems.

That separation made sense when building automation was viewed primarily as maintenance infrastructure.

It makes far less sense now.

As organizations focus on energy management, sustainability goals, asset performance, cybersecurity, and risk mitigation, building data has become valuable beyond the facility department.

Niagara helped create a pathway for that convergence.

Through TCP/IP networking, open APIs, modern web technologies, and flexible integration tools, Niagara enabled building data to flow into enterprise platforms, analytics engines, and cloud applications.

That connection elevated building automation from a maintenance function to a strategic operational capability.

Buildings could now contribute directly to financial performance, sustainability planning, regulatory compliance, enterprise analytics, and operational resilience.

That is a very different role than simply keeping the temperature comfortable.

The Ecosystem That Accelerated Innovation

Niagara's success was not driven by technology alone.

It was accelerated by the ecosystem that formed around it.

OEMs embedded Niagara within equipment. System integrators built custom solutions around it. Developers extended the platform with new applications. Consultants specified Niagara as an open architecture foundation. Building owners began recognizing it as a practical path toward long-term flexibility.

Innovation no longer depended on a single company's roadmap.

It emerged from a global community of developers, manufacturers, integrators, consultants, and owners working around a shared platform.

That is the difference between a product and an ecosystem.

A product solves a defined problem.

An ecosystem keeps expanding what is possible.

Enabling the Smart Building Era

Today's built environment faces expectations that traditional control systems were never designed to meet.

Buildings are now expected to support decarbonization goals, portfolio-wide energy management, demand response, predictive maintenance, cybersecurity governance, AI-driven optimization, digital twin modeling, and occupant experience strategies.

Meeting those expectations requires infrastructure that is open and adaptable.

Closed systems cannot evolve fast enough.

Niagara's architecture allows buildings to integrate emerging technologies without abandoning existing systems. That flexibility matters because buildings do not change all at once. They evolve in layers, over years, often across multiple owners, budgets, contractors, and technology cycles.

A platform that requires everything to be replaced before anything can improve is not a platform built for the real world.

Niagara's strength is that it works with the reality of buildings, not against it.

Closing the Opportunity Gap

The Opportunity Gap was never about a lack of technology.

It was about the absence of a unifying architecture capable of connecting systems, unlocking data, and enabling intelligent operations.

Niagara addressed that gap.

By providing a platform for interoperability, data integration, and enterprise connectivity, Niagara allowed buildings to move beyond isolated automation systems and become connected operational environments.

The result has been a profound shift in how buildings are designed, integrated, and managed.

Systems that once operated independently now work together. Data that was once hidden now informs decisions. Buildings that once reacted to problems can now begin to anticipate them.

That is the real significance of Niagara.

It did not simply connect systems.

It expanded what buildings could become.

A Framework for the Future of Buildings

The built environment is still evolving.

Artificial intelligence, advanced analytics, grid-interactive buildings, digital twins, and intelligent infrastructure are reshaping how buildings operate. These innovations require a foundation built on openness, interoperability, and data accessibility.

Niagara helped establish that foundation.

By solving the integration challenges that once limited building automation, Niagara closed a critical gap between technological capability and operational reality.

It transformed how buildings communicate, how systems collaborate, and how data drives outcomes.

In doing so, Niagara did more than integrate systems.

It helped the built environment move closer to realizing its full potential.

Chapter 7 **The Niagara Doctrine**

Principles That Shaped a Framework

Technology alone rarely reshapes an industry.

Frameworks do.

But frameworks are not built on code alone. They are built on principles—ideas that guide decisions over time, often without being formally documented.

Looking back, the success of the Niagara Framework® was not simply the result of a clever architecture or a capable integration engine. It was the result of a consistent set of beliefs about how building systems should work, how ecosystems should form, and how technology should evolve.

These ideas were rarely written down as formal rules. Yet they influenced partnerships, product direction, and the behavior of the community that formed around the platform.

Taken together, they resemble something more structured.

A doctrine.

Not in the rigid sense—but as a guiding philosophy that helped transform Niagara from a technical framework into a trusted industry platform.

Principle One: Integration Is the Foundation

Buildings are inherently complex environments.

HVAC systems, lighting, security, metering, and countless sensors all operate within the same physical space. Yet historically, they were delivered as independent systems.

Niagara challenged that assumption.

Integration is not a feature.

It is the foundation.

Once systems can communicate, everything else becomes possible. Data becomes accessible. Workflows become coordinated. Operators gain visibility across systems instead of within them.

Buildings stop behaving like collections of equipment.

They begin to function as environments.

This principle remains central to modern smart building architecture—and it is still underestimated.

Principle Two: Openness Enables Innovation

Closed systems slow progress.

When platforms restrict participation, innovation becomes centralized, controlled, and ultimately limited. When platforms are open, innovation becomes distributed.

Niagara embraced openness—not as a marketing position, but as an architectural decision.

Manufacturers could build on the platform. Developers could extend it. Integrators could combine technologies across vendors. Solutions could evolve beyond the boundaries of any single organization.

The result was not just more innovation—but faster innovation.

Because it didn't depend on one roadmap.

It depended on many.

That is how platforms scale beyond their original intent.

Principle Three: Community Creates Trust

Technology does not sustain a platform.

People do.

Over time, Niagara evolved into more than a framework. It became a shared environment where manufacturers, integrators, engineers, and owners worked from a common foundation.

That community didn't form by accident.

It was built through shared events, certifications, training, and real-world collaboration. Knowledge moved between organizations. Experience compounded across projects.

And something important followed:

Trust.

In infrastructure markets, trust matters more than features. Systems are expected to last for decades. Decisions are not easily reversed. Risk is measured in years, not months.

Community accelerates familiarity. Familiarity builds confidence.

And confidence becomes adoption.

Principle Four: Independence Matters

Platforms that favor a single participant eventually lose the ecosystem around them.

Neutrality is not optional.

It is required.

Niagara succeeded in part because it maintained its position as an enabling layer rather than a controlling one. It allowed manufacturers to differentiate, integrators to compete, and developers to innovate—without forcing alignment to a single agenda.

That independence created space.

And that space allowed the ecosystem to grow.

When participants believe the platform is fair, they invest in it. When they believe it is controlled, they hold back.

That difference determines whether a platform expands—or contracts.

Principle Five: Evolution Without Abandonment

Technology changes.

Platforms must change with it.

But there is a difference between evolution and disruption.

Each generation of Niagara introduced new capabilities—web technologies, expanded drivers, improved security, modern development models—while preserving the architectural concepts that made the platform valuable.

That balance matters.

Too much change breaks trust. Too little change breaks relevance.

Niagara navigated both.

It evolved deliberately, without abandoning its core.

And over time, that consistency became one of its strongest advantages.

The Continuing Doctrine

The building automation industry is entering another phase.

Buildings are becoming data environments. Operational intelligence is becoming central. Artificial intelligence is beginning to influence how systems operate.

But the underlying principles have not changed.

Integration still defines capability. Openness still drives innovation. Community still builds trust. Independence still protects the ecosystem. And evolution still determines long-term viability.

Technology will continue to advance.

But platforms built on clear principles tend to endure.

Because when the environment changes—and it always does—principles provide continuity.

And continuity is what turns a framework into infrastructure.

Chapter 8 **The Niagara Movement**

There are technologies. There are products. And then, occasionally, there are movements. The Niagara Framework became a movement—not because it was marketed that way, and not because anyone declared it so, but because it solved a problem the building automation industry had been living with for decades.

For many years, buildings operated in silos. Proprietary systems dominated the landscape. Vendors controlled their ecosystems tightly, and once a system was installed, the owner was effectively locked into that vendor's tools, interfaces, and upgrade path. Integration between systems was complex, expensive, and often fragile. Data remained trapped inside isolated systems. Innovation moved slowly because each ecosystem evolved independently.

Building automation worked, but it did not evolve easily. Niagara changed that.

From Control Systems to Integration Platforms

When Niagara first appeared, its purpose was straightforward: connect disparate building systems. HVAC, lighting, energy meters, security systems, and other operational technologies all spoke different protocols and operated in different software environments. Niagara introduced a framework that could integrate them into a unified architecture.

But what seemed like a technical solution quickly became something more important. Niagara introduced a new way of thinking about building automation—not as a collection of isolated control systems, but as an open integration platform.

Instead of forcing buildings into a single manufacturer's ecosystem, Niagara allowed different systems to communicate. It enabled devices from multiple vendors to coexist. It provided a software framework where integration became part of the architecture rather than an afterthought.

And once integration became easier, something else began to happen. Innovation accelerated. Manufacturers could build Niagara-enabled products. Developers could create drivers and applications. Integrators could design sophisticated multi-system solutions. Engineers could specify architectures that would remain flexible over time. The people responsible for buildings could participate in a way they had not been able to before. The industry began to organize itself around a shared platform. That is when Niagara stopped being simply a technology. It became an ecosystem.

The Rise of the Niagara Ecosystem

Over time, an entire industry community formed around the framework. Original equipment manufacturers began embedding Niagara directly into their products. System integrators built large-scale integration practices around it. Independent developers-based businesses on extending and enhancing it. Consulting engineers began specifying Niagara-based architectures as long-term integration strategies.

Building owners and operators began to recognize the benefits as well. Instead of replacing entire systems when technology evolved, they could extend their existing infrastructure. Integration became scalable. Data became accessible. Systems could evolve over time without forcing disruptive rip-and-replace upgrades.

What emerged was something rare in infrastructure markets: a platform ecosystem. This model was familiar in the IT world, where platforms allow multiple participants to innovate simultaneously – think Microsoft Windows® and the ecosystem it spawned. In the building automation industry—long defined by proprietary control systems—it represented a significant shift.

Niagara provided the foundation. Around it, an ecosystem of manufacturers, integrators, developers, engineers, and building owners began building the next generation of building technology. That is how movements form. Not through declarations, but through shared participation.

From Mechanical Buildings to Data Buildings

Today the industry is entering another transition. Buildings are no longer just mechanical environments. They are becoming data environments.

Every device, sensor, and control point generates information about how a building operates. Equipment performance, energy consumption, environmental conditions, occupancy patterns, and operational events all produce data streams that can be analyzed and acted upon. The focus of building automation is shifting:

- From controls to platforms
- From systems to ecosystems
- From hardware to data
- From features to outcomes

Organizations are increasingly using building data to drive strategic outcomes across entire portfolios. These outcomes include energy optimization, predictive maintenance, operational intelligence, and enterprise-wide performance management.

Technologies such as digital twins, AI-driven analytics, and advanced energy strategies all depend on access to normalized operational data. This evolution requires architectures that are open, scalable, and adaptable. It requires platforms that support innovation rather than restrict it. And once again, the architectural model that Niagara introduced decades ago is proving essential.

The Next Phase: Data Platforms and Digital Infrastructure

The next phase of smart buildings will be defined by how effectively organizations manage and exploit operational data. Buildings now generate enormous volumes of information that can drive:

- Predictive maintenance
- Energy optimization
- Operational intelligence
- Enterprise portfolio management
- Digital twin strategies
- AI-enabled analytics

To support this evolution, building architectures must extend beyond traditional control systems. They must incorporate:

- Edge computing to process data closer to devices
- Data normalization to structure operational information
- Enterprise integration to connect building systems with IT platforms

- Secure connectivity aligned with modern cybersecurity frameworks
- Analytics platforms capable of turning operational data into actionable insight

Niagara remains uniquely positioned within this emerging architecture. Its role as an integration platform allows it to bridge operational technology systems with modern data platforms and enterprise applications. Rather than replacing existing infrastructure, Niagara continues to serve as a connective layer that enables systems to evolve together. This architectural flexibility is one of the reasons the ecosystem continues to grow.

The Expanding Role of System Integrators

As the ecosystem evolves, system integrators have become far more than installers of control systems. They are the architects of digital building infrastructure. The responsibilities of modern integrators increasingly include:

- Designing multi-vendor integration architectures
- Structuring data models and semantic tagging frameworks
- Aligning systems with enterprise cybersecurity policies
- Connecting building infrastructure with enterprise data platforms
- Enabling analytics, AI applications, and portfolio-level intelligence

In many ways, integrators are becoming the equivalent of IT system architects for the built environment. Those who embrace this expanded role will help define the next era of smart buildings. Those who remain focused only on traditional control installation may find themselves left behind as buildings become more data driven. Niagara's ecosystem provides a foundation for integrators to evolve into this role.

A Community That Became a Movement

The success of Niagara has never been about a single product. It has always been about principles.

Openness.
Allowing multiple systems and manufacturers to participate in the same architecture.

Flexibility.
Designing infrastructure that can evolve as technology changes.

Community.
Creating an ecosystem where innovation happens across many participants.

Innovation.
Encouraging developers, manufacturers, and integrators to extend the platform in new directions.

These principles allowed Niagara to grow organically over time. They also allowed the industry itself to mature. The mechanical room still hums. Buildings still need to run. Comfort, safety, and reliability still matter. But the infrastructure that supports those outcomes has evolved dramatically.

Buildings today are no longer isolated control environments. They are connected digital systems.

The Path Forward

Movements rarely announce themselves. They emerge gradually when enough people begin working toward the same set of ideas. Niagara introduced a model for open integration in building automation. Over time, that model enabled a global ecosystem of innovators to participate in shaping the industry.

Today the built environment is entering another period of transformation driven by data, analytics, artificial intelligence, and enterprise connectivity. The foundations that enabled the Niagara ecosystem to remain highly relevant in this new era.

Open platforms. Collaborative ecosystems. Flexible digital infrastructure. The next generation of smart buildings will build upon those ideas. Because movements do not end. They evolve. And the Niagara community continues to move the industry forward.

Chapter 9 **The Ecosystem Effect – How Communities Build Great Brands**

The Ecosystem Effect

There is a truth about technology platforms that is often overlooked:

Technologies do not become movements on their own.

Communities do.

And communities are what turn technologies into enduring brands.

The Niagara Framework did not grow simply because it was technically capable. Many platforms are technically capable. Many claim openness. Many promise interoperability.

What distinguished Niagara was not just what it did.

It was what formed around it.

An ecosystem.

And that ecosystem changed everything.

A Platform Is Only the Beginning

When Niagara was introduced, it solved a real problem.

Buildings were full of isolated systems. Proprietary networks dominated. Integration was difficult, expensive, and fragile. Niagara introduced a framework that could connect disparate systems into a unified environment.

That alone was valuable.

But platforms don't succeed because they exist.

They succeed because people build on them.

Developers extend them. Manufacturers embed them. Integrators deploy them. Consultants specify them. Owners trust them.

When that happens, something important changes.

A platform stops being a product.

It becomes an ecosystem.

Figure 3: The Niagara Ecosystem Wheel

> *The Niagara ecosystem extends beyond a single platform to include integrators, engineers, manufacturers, developers,*

and owners—creating a network that drives adoption, innovation, and long-term trust.

And ecosystems create momentum.

How the Ecosystem Formed

The Niagara ecosystem did not appear overnight.

It emerged gradually—through necessity, opportunity, and alignment of interests.

Manufacturers began embedding Niagara into their devices. Developers built applications that expanded its capabilities. Integrators used it to solve increasingly complex integration challenges. Consultants began specifying Niagara-based architectures because they provided flexibility and long-term viability. Owners started requesting it because it allowed their buildings to evolve instead of stagnate.

Over time, these participants formed something larger than a user base.

They formed a network.

And networks behave differently than individual actors.

Ideas spread faster. Solutions improve faster. Best practices emerge faster.

And most importantly:

Trust spreads faster.

The Power of Network Effects

Economists call this phenomenon a network effect.

As more participants join a platform, its value increases—not incrementally, but multiplicatively.

That dynamic became visible across the Niagara ecosystem.

More manufacturers adopted it, which increased available solutions. More integrators learned it, which increased deployment capacity. More consultants specified it, which increased market demand. More owners requested it, which reinforced adoption.

Each new participant strengthened the platform.

Each new use case expanded its relevance.

The platform didn't just grow.

It accelerated.

A Shared Architecture for Innovation

The ecosystem grew because it had the right foundation.

Niagara did not enforce rigid product boundaries. It provided a shared architectural layer where innovation could occur at multiple levels simultaneously.

Manufacturers innovated at the device level. Developers extended functionality through applications and drivers. Integrators combined systems into cohesive solutions. Owners leveraged the platform to improve operations.

No one had to replace the foundation to contribute.

That is the hallmark of a true platform.

It creates stability at the core—and flexibility at the edges.

That combination allows innovation to scale without breaking the system.

How Communities Build Trust

Brands are often assumed to be built through marketing.

The strongest brands are built through experience—repeated, consistent experience across a community.

When people encounter the same platform across multiple projects, multiple vendors, and multiple environments, something shifts.

Familiarity builds.

Confidence follows.

And eventually, trust becomes institutional.

That is what happened with Niagara.

Engineers saw it specified repeatedly. Integrators encountered it across manufacturers. Owners saw it deployed across portfolios. Developers recognized opportunities to build on it.

The message reinforced itself:

Niagara was not just a tool.

It was becoming infrastructure.

And infrastructure, once trusted, becomes a brand.

The Role of the Community

The strength of the Niagara ecosystem comes from the diversity of its participants.

Manufacturers embed it. Integrators deploy it. Consultants design around it. Developers extend it. Owners depend on it. Educators teach it. Industry groups align around the principles it represents.

Together, they create a feedback loop.

Knowledge spreads. Skills compound. Innovation builds on itself.

Events like Niagara Summits and industry conferences play a critical role—not because of product announcements, but because they reinforce the community. They create shared understanding, shared language, and shared direction.

And when a community aligns, the brand strengthens.

Why Ecosystems Outlast Products

Products change.

Versions evolve. Technologies shift. New tools replace old ones.

But ecosystems endure.

When a platform is supported by thousands of professionals, multiple manufacturers, and a wide range of applications, it becomes resilient. It can adapt to change without losing relevance.

That is one of the reasons Niagara has persisted through multiple waves of technological change—from early integration challenges to enterprise connectivity to data-driven operations.

The ecosystem allowed it to evolve without starting over.

And that ability to evolve is what separates temporary solutions from enduring platforms.

Why This Matters for the Future of Buildings

The built environment is becoming more complex, not less.

Buildings now integrate multiple systems, devices, applications, and data sources. No single vendor can own all of it. No single product can manage it alone.

The future will be built on ecosystems.

Platforms that allow multiple participants to contribute, innovate, and adapt over time.

That is the environment Niagara helped create.

And that model will only become more important as buildings become more connected, more intelligent, and increasingly autonomous.

The Lesson for the Industry

The Niagara story offers a clear lesson.

Great brands are not built solely through technology.

They are built through participation.

Through ecosystems where many organizations contribute value. Through communities where knowledge accumulates. Through platforms that enable innovation instead of restricting it.

When those conditions exist, a predictable progression follows:

Technology becomes trusted.
Trust becomes adoption.
Adoption becomes infrastructure.
Infrastructure becomes identity.

That is the ecosystem effect.

And it is why the Niagara Framework became more than a product.

It became a movement.

Chapter 10 **From Framework to Flagship — How Niagara Became a Brand**

The Niagara Framework did not set out to become a brand.

It began as something far more practical—an integration layer designed to connect building systems that were never meant to communicate with one another. HVAC, lighting, access control, metering, and fire systems each existed in separate silos, speaking different protocols and managed through proprietary tools.

Niagara solved a technical problem. Over time, however, something more significant happened. It became a brand—not through advertising campaigns or aggressive promotion, but through discipline. Consistent decisions about positioning, evolution, and user experience created something rare: recognition without promotion, trust without enforcement, and adoption without coercion.

A Market Without Platform Identity

Before Niagara, building automation was dominated by large manufacturers—Johnson Controls, Honeywell, Siemens, and Schneider Electric.

These companies had strong corporate brands, but the layer that connected systems did not. Each manufacturer operated a vertically integrated stack of controllers, software, graphics, networking, and service infrastructure. Selecting a vendor meant committing to that ecosystem. Changing vendors meant starting over.

The industry had products, equipment, and recognizable manufacturers. What it lacked was an independent platform identity. Niagara changed that by making the integration layer visible—and by giving it a name.

The Power of a Name

"Niagara" was not a technical label, an acronym, or a feature description. It created imagery. A waterfall connects systems of water across geography; Niagara connected systems across buildings. The metaphor was subtle, but effective.

In an industry dominated by model numbers and abbreviations, the name stood out. It was memorable, and that matters more than most people realize. In technical markets, recall is the first step toward preference—and preference is where brand begins.

More Than a Product

One of the most important decisions behind Niagara was how it was positioned. It was not introduced as a product, but as a framework.

That distinction is not semantic—it is strategic. Products are replaced. Frameworks persist.

A product delivers a defined set of capabilities. A framework creates an environment where capabilities can continue to evolve. It provides structure, services, and rules that allow others to build solutions on top of it. That is what the Niagara Framework became for the built environment.

It did not compete with systems. It connected them. And because of that, it outlived them.

Enabling Instead of Controlling

Most platforms are built to capture value. Niagara was built to distribute it.

Manufacturers embedded it into their products. Integrators built businesses around it. Developers extended it. Consultants specified it. Owners requested it. The platform expanded horizontally across the industry rather than vertically under a single organization.

That created a fundamentally different dynamic. Niagara did not dominate the ecosystem—it enabled it. Instead of creating dependency, it created loyalty. That difference is subtle, but it is foundational to how the brand developed.

Consistency Over Time

Technology platforms often fracture as they evolve. Version changes break compatibility, development models shift, and users are forced to start over. When that happens, ecosystems weaken.

Niagara avoided that outcome. From R2 to AX to Niagara 4, the platform evolved significantly, but it never became unrecognizable. The core experience remained consistent. Workbench stayed central. The programming model evolved without being discarded.

The ecosystem moved forward together. That continuity matters, because consistency builds trust—and trust compounds over time.

When the Market Starts Saying Your Name

At some point, Niagara stopped being described and started being referenced.

It appeared in specifications, RFPs, job descriptions, and training programs. It became part of everyday language within the industry. Engineers and owners began asking a simple question: "Is it Niagara?"

That question was not about features. It was about expectations—open, interoperable, flexible, extensible, and future-ready. When a name becomes shorthand for a set of assumptions, it has crossed a threshold. It is no longer just a technology. It is a brand.

Independence as Identity

When Honeywell acquired Tridium, many expected Niagara to become a captive technology. That would have aligned with typical industry behavior.

Instead, Niagara remained open. OEM partnerships continued. Integrators across multiple ecosystems kept using it. That neutrality became part of its identity.

Niagara came to be seen as the independent integration layer—the connective tissue of the smart building. Some described it as the "Switzerland of integration." That perception mattered, because independence signals trust in markets where long-term decisions carry significant risk.

The Role of Community

No brand is built in isolation. Niagara grew because a professional community formed around it.

Developers learned it. Integrators built careers on it. Consultants specified it. Owners demanded it. Training programs and certifications reinforced it. Over time, people invested not just financially, but professionally.

When individuals build their careers around a platform, they do more than use it—they advocate for it. That is when adoption turns into momentum, and momentum turns into brand strength.

Adapting Without Losing Identity

Every platform is eventually tested. Niagara faced multiple waves of change, including cloud computing, IoT, advanced analytics, and artificial intelligence.

Each wave raised the same question: would it become obsolete?

It did not. It adapted. Edge capabilities expanded, APIs matured, security strengthened, and cloud integration improved. The platform evolved from an integration layer into a broader operational technology platform.

What it did not do was abandon its core identity. That balance—evolution without fragmentation—is rare, and it is one of the reasons Niagara endured.

What Actually Built the Brand

Niagara did not become a brand because it set out to be one. It became a brand because it delivered consistently over time.

It worked.
People built businesses on it.
It became a recognized descriptor in the market.
Others promoted it voluntarily.
It outlived multiple generations of technology.

Most products fade as versions change. Niagara did not. It became something larger than its versions.

Discipline Without Arrogance

One of the most overlooked aspects of Niagara's success was restraint.

It did not overpromise. It did not attempt to dominate the ecosystem. Instead, it allowed others to succeed within it. Manufacturers differentiated their products. Integrators built their own brands. Consultants maintained independence. Partners created value.

Niagara did not try to own that value. It enabled it.

That approach built something that cannot be forced or manufactured—trust.

The Brand That Emerged

Today, Niagara represents more than software. It represents a philosophy of building technology: openness, interoperability, continuity, flexibility, and professional credibility.

In an industry historically defined by products, Niagara demonstrated something different. A framework can become a brand.

Technology creates capability. Consistency builds trust. Community creates momentum. Longevity creates credibility. Brand emerges when those elements align and are reinforced over time.

Niagara did not become a brand because it declared itself one. It became a brand because the industry decided it was one.

And when the market begins using your name as shorthand for value, you are no longer just a framework.

You are a flagship.

Chapter 11 **Niagara Joined the Great Brands**

In every major industry, a small number of companies cross an invisible line.

They stop being vendors.
They stop being options.
They become infrastructure.

When that happens, the nature of competition changes. You are no longer compared feature-to-feature. You become the baseline against which others are measured.

In computing, that role has been held by companies like Intel and IBM. In networking, Cisco. In enterprise software, Oracle and Salesforce. In industrial systems, General Electric. In digital workflows, Adobe.

Each operates in a different domain.

But they share a common outcome:

They became foundational to how their industries function.

And in the built environment, the Niagara Framework has moved into that same category.

What the Great B2B Brands Have in Common

These companies did not arrive at that position by accident.

They share a consistent set of characteristics.

They operate at the core of business operations, not at the edges. They are trusted to run mission-critical systems. They are supported by ecosystems of partners, integrators, and trained professionals. They persist across decades of technological change. And they focus on outcomes rather than individual products.

Intel didn't just sell chips—it became the computing engine behind the modern world.

IBM didn't just sell hardware—it became the trusted backbone for enterprise systems.

Cisco didn't just sell networking gear—it became the infrastructure of the internet.

Salesforce didn't just sell software—it redefined how organizations manage relationships and workflows.

In each case, the brand became associated with reliability, continuity, and scale.

That is what separates strong brands from foundational ones.

From Product to Platform

Most technology companies begin by selling products.

The strongest B2B brands evolve into platforms.

That transition is where real differentiation happens.

Products solve defined problems. Platforms create environments where many problems can be solved over time.

Niagara followed this same path.

In its early years, it was often described as an integration tool or a BAS front end. That description was technically accurate—but strategically incomplete.

What was actually happening was more significant.

Niagara was becoming a runtime environment, a development framework, a data normalization layer, and an integration fabric.

It wasn't just controlling buildings.

It was orchestrating systems.

Over time, Niagara stopped being evaluated as a tool and started being trusted as a foundation.

And that distinction changes everything.

The Power of Interoperability

One of the defining traits of enduring B2B brands is neutrality.

They do not win by excluding others.

They win by enabling participation.

Niagara embraced that philosophy early.

While many competitors built closed ecosystems, Niagara built bridges. It integrated protocols, systems, and applications instead of replacing them. It asked a different question—not "How do we take over?" but "How do we make everything work together?"

That decision lowered risk.

Owners could adopt Niagara without abandoning existing investments. Integrators could work across vendors. Consultants could specify solutions without locking clients into a single path.

Lower risk increased adoption.

Adoption created momentum.

Momentum built brand.

Ecosystem Over Ownership

No great platform dominates alone.

Niagara succeeded because it did not try to vertically integrate the entire stack. Instead, it cultivated an ecosystem of OEM partners, system integrators, developers, consultants, and owners.

That ecosystem became self-reinforcing.

As more professionals learned Niagara, it became safer to specify. As more firms standardized on it, it became harder to displace. As more tools were built around it, it became more valuable.

That is how platforms become defensible.

Not through patents.

Through participation.

Where Great Brands Live

Another defining trait of these brands is where they operate.

They exist in the most sensitive parts of organizations.

IBM in financial systems.
Oracle in enterprise data.
Cisco in network infrastructure.
GE in critical industrial systems.

Niagara occupies a similar position.

It lives inside buildings—the physical environments where organizations operate. If it fails, operations are affected. Comfort, safety, energy performance, and system coordination all depend on it.

That reality changes how customers think.

Niagara is not just software.

It is operational infrastructure.

And infrastructure demands stability, predictability, and long-term support.

From early on, Niagara emphasized backward compatibility, upgrade paths, and lifecycle continuity. It favored durability over disruption.

That discipline reinforced trust.

The Brand DNA of Niagara

Over time, Niagara developed a recognizable identity—its own internal logic for how it behaves in the market.

It is defined by openness, integrating rather than replacing systems. It maintains neutrality, enabling rather than competing with its ecosystem. It normalizes complexity, turning fragmented systems into usable data. It scales across buildings, campuses, and global portfolios. And it grows through community, aligning its success with the success of its partners.

These are not marketing messages.

They are operating principles.

And they mirror the same patterns seen in the strongest B2B brands across industries.

Becoming the Default

At a certain point, strong brands reach a different stage.

They stop being evaluated individually.

They become the reference point.

When that happens, competition shifts. Instead of comparing features, buyers ask a different question:

"How does this compare to the standard?"

Niagara has crossed that threshold.

It has become the baseline for integration in building automation.

That position is not declared.

It is granted—by the market.

Conclusion: When a Framework Becomes a Promise

No great brand succeeds without trust.

And trust is not created in a moment. It is built over time—through consistency, openness, reliability, and shared success.

Niagara's breakthrough was not that it competed as a better control system.

It was that it stopped competing at the product level altogether.

It became a universal framework.

It made other companies more capable. It made integrators more effective. It made OEMs more competitive. It gave owners confidence that their buildings could evolve over time.

It turned fragmentation into coherence.

And in doing so, it joined a small group of platforms that define their industries—not by controlling them, but by enabling them.

That is the mark of a true infrastructure brand.

And that is Niagara's legacy.

Chapter 12 **Why Niagara Has Endurance**

In every technology-driven industry, a familiar pattern appears.

A breakthrough platform emerges. It reshapes expectations. It becomes a reference point. And almost immediately, challengers arrive—each promising to be faster, simpler, more modern, or more complete.

Building automation has followed this pattern with remarkable consistency.

For more than two decades, dozens of platforms, architectures, and "next generation" solutions have attempted to replace or outflank the Niagara Framework. Some arrived with enormous corporate backing. Others with bold technical visions. Many were introduced with confidence that this time the industry would finally move on.

And yet, decades later—with more than a million deployed instances globally—Niagara remains a foundational platform.

Not because innovation stopped.
Not because alternatives never appeared.
Not because customers lacked choice.

Niagara endured because it was designed for endurance.

The Industry's Endless Reset Cycle

The building automation industry has always been searching for a definitive architecture—a unified interface, a universal data model, a standard operating environment for buildings.

Each new wave of technology revives that ambition.

Over time, the industry has cycled through proprietary front ends, enterprise energy platforms, cloud-first architectures, IoT hubs, digital twins, and AI-driven control layers. Each promised to solve what came before.

And each, in some way, underestimated the same thing:

Buildings are not clean systems.

They are layered environments, assembled over decades, containing equipment from multiple vendors, communicating across different protocols, installed by different contractors, and operated under changing ownership structures.

Any platform that assumes a clean slate begins with a disadvantage.

Niagara never made that assumption.

Why So Many Challengers Fell Short

Most attempts to replace Niagara started with a reasonable premise—and then pushed it too far.

Some believed vertical integration was the answer. If they controlled hardware, software, networking, tools, and services, they could lock in customers and dominate the lifecycle.

Inside those ecosystems, the systems often worked well.

Outside them, they struggled.

Buildings rarely remain loyal to a single vendor. Equipment changes. ownership changes. integrators change. Contracts change. Closed systems degrade as the environment around them evolves.

Other challengers pursued simplicity. Their goal was faster deployment, cleaner interfaces, and turnkey solutions.

In controlled environments, many succeeded.

At scale, they often struggled.

What works in a pilot does not always work across diverse portfolios.

Then came the IT-driven challengers—cloud-native platforms, API-first ecosystems, containerized architectures. These positioned Niagara as legacy.

But many failed to reconcile modern IT assumptions with operational reality.

Buildings do not operate like cloud applications. Connectivity is not guaranteed. Systems must run continuously. Failures have physical consequences.

Modern architecture alone does not create resilience.

Buildings Are Not Apps

This is where many challengers miscalculated.

Buildings are long-lived physical assets. Equipment installed today may remain in service for decades. Systems are upgraded in phases. ownership changes. Technologies accumulate rather than reset.

Protocols from the 1990s coexist with modern APIs. Legacy controllers operate alongside cloud analytics platforms.

The building is not a clean environment.

It is a timeline.

Platforms that require standardization before they can operate rarely succeed in that reality.

Niagara was built for it.

Architecture Designed for Longevity

Niagara's endurance is rooted in architectural decisions.

From the beginning, it avoided dependence on a single protocol. Instead, it embraced neutrality—supporting BACnet®, LONWORKS®, Modbus®, and others as the industry evolved.

This prevented technological lock-in and ensured relevance across generations.

Its modular architecture reinforced that flexibility. Connectivity, logic, visualization, and data handling were designed as components, not a monolith. New capabilities could be added without rebuilding the platform.

Equally important was compatibility.

Niagara respected the installed base. Upgrades preserved prior investment. Users were not punished for staying.

In infrastructure markets, that principle matters more than feature velocity.

The Power of Community

Many competitors built products.

Niagara built a community.

Developers, integrators, OEMs, consultants, and owners became participants in a shared ecosystem. Innovation no longer depended on a single roadmap—it emerged from a network.

That network became a strategic advantage.

Integrators, in particular, played a defining role. They design systems, install them, maintain them, and troubleshoot them over decades.

Niagara empowered them.

And they, in turn, advocated for it.

Platforms that bypass integrators lose influence.

Platforms that empower them gain momentum.

Adaptation Without Abandonment

Every platform faces disruption.

Niagara faced many: web technologies, cybersecurity demands, IT governance, cloud integration, edge computing, analytics, and now artificial intelligence.

Each shift required evolution.

Niagara absorbed these changes without abandoning its core.

It expanded rather than replaced. It extended rather than reset.

That balance—between continuity and progress—is rare.

Some platforms change so aggressively they fracture their user base. Others resist change and become irrelevant.

Niagara avoided both.

Endurance as Strategy

Niagara's longevity is not just technical.

It is strategic.

Over time, it established a consistent identity:

Stable, but not stagnant.
Open, but not chaotic.
Neutral, but not passive.

Flexible, but not fragile.
Trusted, but not complacent.

Those attributes were not marketing language.

They were reinforced through architecture, ecosystem decisions, and long-term behavior.

Consistency became credibility.

Credibility became trust.

And trust became the foundation of the brand.

Why Niagara Endures

Many platforms have tried to replace Niagara—cloud-native startups, vertical stacks, IoT disruptors, AI-native systems.

Some introduced meaningful innovation. Some continue to contribute value.

But few were designed for the reality of buildings.

Niagara was.

It embraced protocol diversity. It supported incremental evolution. It protected existing investment. It empowered a broad ecosystem.

And it evolved without abandoning its identity.

That is not just engineering discipline.

It is brand maturity.

Conclusion

Niagara did not endure because it was the most aggressive platform.

It endured because it was the most aligned with reality.

Buildings are complex, long-lived, and constantly evolving systems. Platforms that succeed in that environment must prioritize continuity, flexibility, and trust over short-term advantage.

Niagara did that.

And over time, that discipline created something larger than a technology platform.

It created a brand defined by endurance.

Chapter 13 **Lessons for the Industry — What Can We Learn from Niagara**

Lessons for the Industry

For more than two decades, the Niagara Framework ecosystem has served as a real-world case study in how to build, grow, and sustain a technology platform in a conservative, fragmented, and engineering-driven industry. While many building automation technologies have risen and faded, Niagara has endured—and in many ways helped define the modern smart building landscape.

That endurance was not accidental. It was the result of architectural discipline, ecosystem thinking, and a long-term commitment to trust. When viewed through that lens, Niagara's evolution reveals a set of consistent principles—lessons that extend beyond building automation into any industry attempting to build durable platforms.

Build Frameworks, Not Just Products

From its earliest days, Niagara was conceived as a framework rather than a single-purpose product. That distinction proved foundational. Most building automation solutions are designed to solve a defined problem—control a system, manage a device, or provide a specific interface. As those problems evolve, the solutions are often replaced.

Niagara approached the problem differently. Instead of solving for a single use case, it created an environment where multiple systems, data sources, and workflows could coexist and evolve together. By focusing on integration, abstraction, and extensibility, it positioned itself not as a point solution, but as a long-term foundation.

This is the core lesson: platforms that endure are not built around today's requirements alone. They are designed to accommodate tomorrow's uncertainty. If a solution cannot extend beyond its initial use case, it is not a framework—it is software with a limited lifespan.

Make Openness Structural

Openness is frequently claimed in technology markets, but rarely implemented in a meaningful way. In Niagara's case, openness was not a feature layered on top of the platform—it was embedded into its architecture. Protocol neutrality, multi-vendor support, extensibility, and third-party development were structural decisions made early and reinforced over time.

This approach directly addressed one of the industry's most persistent concerns: vendor lock-in. By reducing dependency on any single manufacturer, Niagara allowed owners to protect long-term investments, integrators to maintain flexibility, and developers to innovate without restriction.

The broader lesson is straightforward but often ignored: openness reduces risk. When customers perceive lower risk, they are more willing to adopt, invest, and expand. True openness, however, requires discipline. It often means surrendering short-term control in exchange for long-term ecosystem growth.

Empower Partners Instead of Competing with Them

A defining characteristic of Niagara's strategy was its relationship with partners. Rather than attempting to dominate every layer of the value chain, it enabled others to build successful businesses on top of the platform. OEMs, integrators, and solution providers were not overshadowed—they were amplified.

This decision created a multiplier effect. As more organizations succeeded within the ecosystem, the platform itself became more valuable. Integrators developed expertise, manufacturers differentiated their offerings, and consultants gained confidence in specifying Niagara-based architectures.

Many platforms fail at this stage. As they grow, they begin to compete with their own partners, capturing more of the value chain and, in the process, weakening the ecosystem that supported their growth. Niagara avoided that trap by maintaining a clear boundary: it enabled the ecosystem rather than competing with it.

The result was not dependency, but alignment.

Invest in Community Before Monetizing It

Niagara's growth cannot be fully understood without recognizing the role of its community. Conferences, certification programs, training initiatives, developer forums, and partner networks created an environment where knowledge circulated rapidly and expertise compounded over time.

More importantly, the community created identity. Professionals did not simply use Niagara—they became associated with it. Careers were built on it. Reputations were formed within its ecosystem. Over time, this created a level of loyalty that extended beyond technical preference.

This highlights a critical principle: communities create switching costs that no contract can enforce. When individuals invest their time, skills, and professional identity into a platform, the decision to move away becomes significantly more complex.

Platforms that treat community as an afterthought rarely achieve this level of durability.

Consistency Builds Trust at Scale

Niagara evolved continuously, but it did so with restraint. Users came to expect that upgrades would be manageable, that core principles would remain intact, and that investments in both technology and training would be preserved. This predictability enabled long-term planning and reduced operational risk.

In environments such as hospitals, campuses, and critical infrastructure, stability is not optional. It is essential. While innovation is important, it must be delivered in a way that organizations can absorb without disruption.

Niagara demonstrated that progress and stability are not mutually exclusive. By maintaining continuity while introducing new capabilities, it allowed the ecosystem to evolve without fragmentation.

Over time, that consistency became a defining characteristic of the platform's brand.

Balance Engineering with Governance

Technical excellence was necessary for Niagara's success, but it was not sufficient. The platform's durability was reinforced by disciplined governance—structured licensing, partner programs, training frameworks, and consistent messaging.

Many technically impressive platforms fail because they overlook this dimension. They focus on capability but neglect sustainability. Without governance, even strong technology becomes difficult to scale and support over time.

The lesson here is clear: engineering creates potential, but governance preserves it. The platforms that endure are those that treat both with equal importance.

Shape the Language of the Industry

One of Niagara's most subtle advantages was its influence on how the industry thinks. Over time, its terminology—stations, points, histories, alarms, and networks—became part of the shared vocabulary of building automation.

This is more significant than it appears. Language shapes mental models, and mental models shape design decisions. When professionals begin to think in terms defined by a platform, that platform becomes embedded in how solutions are conceived and implemented.

In effect, Niagara did not just provide tools. It provided a way of thinking about systems.

Platforms that achieve this level of influence move beyond visibility. They become assumed.

Avoid Brand Arrogance

Despite its influence, Niagara maintained a collaborative posture. It did not position itself as the only viable solution, nor did it attempt to dominate the ecosystem it helped create.

This restraint contributed to its longevity. Markets tend to resist platforms that overreach, particularly in industries where multiple stakeholders must collaborate over long periods.

Confidence encourages participation. Arrogance discourages it.

Niagara's ability to maintain this balance allowed it to lead without alienating the very partners it depended on.

Design for Decades, Not Quarters

Niagara's roadmap was not driven by short-term trends. Over time, it adapted to major technological shifts—web technologies, cloud integration, analytics, and emerging artificial intelligence—without abandoning its core architecture.

This approach required patience. It also required discipline. Rather than chasing every emerging trend, Niagara evaluated how new technologies could be integrated into its existing framework.

The result was continuity. While other platforms were replaced or restructured, Niagara expanded.

The lesson is simple: trends are temporary. Infrastructure is long-lived. Platforms designed for longevity prioritize integration over reinvention.

Treat Brand as Strategic Capital

Niagara's brand was not built through marketing campaigns alone. It was built through consistent behavior. Over time, the platform became associated with interoperability, openness, and reliability—not because those attributes were promoted, but because they were repeatedly demonstrated.

Every release, every training program, and every partner interaction reinforced the same expectations. That consistency transformed brand from messaging into experience.

Brand, in this sense, is cumulative. It reflects how a platform behaves over time.

Organizations that treat brand as an afterthought often struggle to maintain trust. Those that treat it as a strategic asset build resilience.

What the Industry Must Learn

As building automation converges with IT systems, cloud platforms, and data-driven operations, the pressure to define the next generation of platforms will intensify. Many solutions will emerge, each promising to address gaps in the current landscape.

The platforms that endure will not simply introduce new features. They will internalize the principles demonstrated by Niagara—framework thinking, structural openness, partner alignment, community investment, consistency, governance, disciplined language, humility, long-term design, and brand as behavior.

These are not historical observations. They are requirements for durability.

Closing Reflection

Niagara's greatest contribution is not measured solely in deployment scale or market presence. Its deeper impact is the shift in how the industry understands success.

It demonstrated that integration matters more than individual devices, that ecosystems outperform isolated solutions, and that trust outlasts innovation cycles. It showed that frameworks can outlive products—and that platforms built with discipline can shape entire industries.

For organizations looking ahead, the lesson is not to replicate Niagara's technology, but to understand its philosophy.

Because in the end, Niagara did not just build a platform.

It built a blueprint for endurance.

Chapter 14 **What Niagara Means for the People Who Actually Build and Run Buildings**

If brand were simply a marketing concept, it would not matter much in the building automation industry.

Buildings are not purchased like consumer products. No one walks into a store and selects a building automation system off the shelf. These systems are designed, specified, engineered, integrated, installed, and operated over decades. They are part of critical infrastructure.

And yet, brand matters here as much as anywhere—perhaps more.

Because in infrastructure industries, brand becomes shorthand for something deeper: trust, confidence, longevity, and the belief that the platform chosen today will still serve its purpose tomorrow.

The Niagara Framework® did not become a brand through marketing alone. Marketing helped, but it was not the reason. Niagara became a brand because the people who design, install, and operate buildings discovered that it consistently solved real problems and delivered outcomes over time.

This chapter is for those people—system integrators, consulting engineers, and building owners and operators—because platforms only endure when the people who rely on them continue to believe in them.

The System Integrator Perspective

The People Who Make Buildings Work

System integrators are the ones who make modern buildings function. They are in mechanical rooms early in the morning, opening control panels, troubleshooting networks, integrating legacy systems, and ensuring that everything operates as intended.

For decades, integrators worked within rigid vendor ecosystems. Choosing a manufacturer meant committing to its controllers, software, tools, licensing structure, and limitations. The business model was straightforward: align with a vendor, build expertise within that ecosystem, and operate within its boundaries.

That model worked—but it came with constraints. It limited flexibility, slowed innovation, and often prevented integrators from fully responding to what their customers actually needed.

Niagara introduced a different model. It gave integrators a framework instead of a silo—a platform that could connect systems rather than replace them, and one that allowed multiple manufacturers to coexist within a single environment.

The impact was significant. Integrators were no longer forced to think in terms of one system per building. They could think in terms of one framework connecting everything. That shift expanded service opportunities, enabled more sophisticated solutions, and allowed integrators to evolve from installers of proprietary systems into providers of integrated solutions.

Just as importantly, Niagara respected their expertise. It did not replace the integrator's role—it expanded it. That is a major reason it earned long-term loyalty within that community.

The Consulting Engineer Perspective

Designing Buildings That Must Last Decades

Consulting engineers operate on a different timeline. The systems they design are expected to perform for decades, often across multiple ownership cycles, tenant changes, and technological shifts.

When an engineer writes a specification, they are not selecting technology for today—they are defining infrastructure that must remain viable in the future. Historically, this meant choosing a vendor ecosystem, a decision that carried inherent risk. If that vendor's technology became obsolete or failed to evolve, flexibility was limited and options were constrained.

The introduction of open protocols such as BACnet and Modbus improved interoperability, but protocols alone did not solve the architectural problem.

What engineers needed was a unifying layer—something that allowed systems to coexist without being locked into a single path.

Niagara provided that layer. It functioned as a neutral integration framework, independent of any single manufacturer or system type. It allowed engineers to design buildings that could evolve, integrate new technologies, and adapt over time without requiring wholesale replacement.

That flexibility is not just a technical advantage—it is a form of risk mitigation. And for engineers responsible for long-term system performance, reducing risk is one of the most valuable outcomes they can deliver.

The Building Owner Perspective

Buildings as Long-Term Investments

For building owners and operators, technology decisions are rarely about features. They are about outcomes—operating costs, energy performance, risk exposure, tenant experience, and long-term asset value.

The challenge is that those outcomes depend on infrastructure that must remain flexible over time. Many owners have experienced the consequences of vendor lock-in: systems that cannot integrate, platforms that cannot scale, and technologies that cannot adapt. When that happens, the cost is not just financial—it is operational.

Buildings become harder to manage. Data becomes fragmented. Innovation slows.

Niagara shifted that dynamic by giving owners something they rarely had before: architectural independence.

With Niagara, owners could integrate systems from multiple vendors, add analytics platforms, connect energy systems, and expand across portfolios without starting over. They could evolve their building strategy incrementally instead of through disruptive replacement cycles.

That capability aligns with a simple but powerful principle:

Buildings should serve their owners—not the other way around.

When Infrastructure Becomes Trusted

The most powerful brands in infrastructure rarely feel like brands. They are not constantly visible. They do not demand attention. They simply become part of the environment.

Niagara followed that path. It became trusted because it consistently worked, evolved without abandoning its core principles, and allowed innovation to happen around it rather than forcing it through a single channel.

Over time, an ecosystem formed—one that included integrators, engineers, manufacturers, developers, and owners. That ecosystem created something that marketing alone cannot produce: shared ownership.

When a community begins to feel ownership of a platform, that platform becomes more than infrastructure.

It becomes part of professional identity.

A Final Message to the People Who Build the Industry

If you are a system integrator, engineer, or building owner, you are not just participating in the industry—you are shaping it.

Every system you design, every building you integrate, and every platform you choose influences how buildings will operate for decades.

The story of Niagara is not just about technology. It is about a shift in how the industry thinks—from proprietary systems to open frameworks, from isolated buildings to connected infrastructure, and from control systems to platforms.

As the built environment continues to evolve toward intelligent—and eventually autonomous—systems, the importance of adaptable, open frameworks will only increase. Buildings are no longer purely mechanical environments. They are digital systems that require architectures capable of supporting continuous change.

Niagara became one of the architectures the industry depends on.

Not because it tried to dominate the industry, but because it enabled the industry to grow.

And that, ultimately, is why the people who build and operate buildings continue to trust it.

Chapter 15 **What Brand Really Is**

Brand is not a logo. It is not a tagline, a website, or a trade show booth. Those are expressions of brand—tools used to communicate it—but they are not the brand itself. Brand is accumulated behavior. It is the memory of how an organization performs over time: the pattern of its decisions, the consistency of its philosophy, the predictability of its roadmap, and the integrity of its support.

For much of the building automation industry's history, brand barely registered. Companies competed on hardware specifications, protocol support, installed base, and price. Success was measured in bid wins, point counts, and service contracts. Marketing, when it existed, was largely tactical—brochures, product sheets, and trade show presence. Brand was treated as cosmetic.

Over time, however, something changed. Certain companies began winning even when their products were similar. Certain platforms became trusted defaults. Certain names carried weight before a proposal was even opened. The industry did not suddenly become emotional—it became experiential. As buildings grew more complex, more connected, and more risk-sensitive, decisions were no longer made on specifications alone. They were shaped by accumulated experience. In that environment, brand stopped being surface-level. It became part of how decisions are made.

To understand this shift, it helps to be precise about what brand is not. Brand is not your logo, your color palette, your slogan, or your messaging. Those elements can be redesigned without changing the underlying reality. A company can refresh its identity and still deliver the same inconsistent experience. Brand exists beneath those elements. It lives in the minds of customers, partners, and employees, shaped by thousands of interactions over time. Your brand is not what you say it is. It is what people have learned to expect from you.

At its core, brand is a pattern of perception formed through repeated experience. Every interaction contributes to that pattern: sales conversations, support responses, product releases, training sessions, installations, integrations, outages, upgrades, invoices, and promises kept or broken. Over time, those experiences accumulate. When someone hears your name, something forms instantly in their mind—dependable, difficult, safe, risky, forward-looking, or stuck in the past. That reaction is your brand.

In building automation, those perceptions carry weight because every decision involves risk. Owners risk capital. Integrators risk reputation. Engineers risk credibility. Operators risk uptime. IT teams risk security. Executives risk accountability. No one buys controls in isolation—they buy consequences. Will the system still be supported in ten years? Will updates break integrations? Will the company still be around? Will the team be able to operate it effectively? These are not feature questions. They are trust questions.

This is where brand becomes infrastructure. In a risk-sensitive industry, brand functions as a form of confidence. It reduces uncertainty and allows decision-makers to move forward with conviction. When two platforms appear similar on paper, the one with the stronger track record—the one that people believe will hold up over time—wins. Not because it is marketed better, but because it is trusted more.

Brand also enables what might be called confidence transfer. When an integrator recommends a platform, they are transferring their credibility to it. When an engineer specifies a system, they are attaching their professional reputation to that decision. When an owner standardizes across a portfolio, they are committing long-term operational performance to that choice. A strong brand supports that transfer. It gives people the confidence to recommend, specify, and invest without hesitation. A weak brand introduces doubt, and doubt slows or kills decisions.

Every brand carries a promise, whether explicit or implied. In building automation, that promise is not about features—it is about consistency over time. Not just at startup or commissioning, but years into operation. A single successful project does not build a brand. Consistency across many projects does. Likewise, a single failure may not destroy a brand, but patterns of inconsistency will. Brand is not defined by peak performance. It is defined by reliability over time.

From the outside, brand often appears to be messaging. From the inside, it is alignment. In strong organizations, product development, support, sales, and leadership operate from a shared understanding of what the company stands for and how it behaves. Decisions reinforce one another. In weaker organizations, those elements drift apart. Sales overpromises. Engineering underdelivers. Support improvises. Leadership changes direction. Customers feel that inconsistency immediately, and it erodes trust faster than any competitor can.

Brand is also economic. It affects how business gets done. Organizations with strong brands tend to move faster. Sales cycles shorten. Win rates improve. Pricing pressure decreases. Partners stay longer. Customers return. The reason is simple: trust reduces friction. When customers believe in a platform, they spend less time second-guessing the decision and more time implementing it.

In this industry, customers do not remember specification sheets. They remember moments. Who answered the phone at midnight. Who resolved the integration that no one else could fix. Who stood behind a system when something failed. Who disappeared after installation. Those moments become stories. Stories become reputation. And reputation becomes brand.

Importantly, brand is built internally before it is expressed externally. Many organizations attempt to build brand from the outside in—starting with messaging, design, or campaigns. Strong brands are built the opposite way. They begin with culture, technical philosophy, product discipline, and long-term decision-making. External identity reflects internal reality. Without that foundation, branding becomes decoration. With it, branding becomes amplification.

Over time, brand shapes strategy. It influences product roadmaps, partnership decisions, pricing models, and market positioning. Every strategic decision either reinforces the brand or weakens it. Short-term gains that conflict with long-term expectations leave a mark. Customers notice. Partners notice. The market remembers.

Technologies will change. Protocols will evolve. Product lines will rise and fall. What endures is the pattern behind them—the consistency of behavior that customers come to rely on. Strong brands outlive products because they represent something deeper: a predictable way of working over time.

In a fragmented industry like building automation, that predictability matters. Modern buildings are complex environments involving multiple vendors, systems, and stakeholders. Buyers look for stability within that complexity. Brand provides that anchor. It signals identity, continuity, and reliability. Without it, companies become interchangeable—and interchangeable companies compete primarily on price.

Price competition is a losing position in the long run. Brand is the alternative. It allows organizations to differentiate not just on what they deliver, but on how consistently they deliver it.

At its simplest, brand comes down to a single idea: the collective belief that you will do what you say you will do—consistently, competently, and with integrity over time. That belief is built slowly and reinforced continuously. In an industry where systems last decades and relationships span careers, it may be the most valuable asset a company can build.

Chapter 16 **The Brand Promise**

In every industry, strong brands make promises.

In building automation, they make commitments.

They commit to systems that work when failure is not an option.
They commit to performance that affects energy, comfort, safety, and asset value.
They commit to relationships that often outlast the teams who installed the system.

This is not marketing.

This is obligation.

What a Brand Promise Really Is

A brand promise is not a slogan.

It is the expectation you set—and the behavior you repeat.

It answers a simple question:

"If I choose you, what can I rely on over time?"

Not just at startup.
Not just during commissioning.
But five, ten, even twenty years into operation.

That promise shows up in very practical ways:

How your system performs under stress.
How your roadmap evolves.
How you respond when something breaks.
How you treat partners.
How predictable your decisions are.

It is an invisible contract.

And once the market believes it, you don't get to selectively honor it.

Why It Matters More in This Industry

In most industries, switching is inconvenient.

In building automation, switching is disruptive, expensive, and often avoided entirely.

Once a system is installed, it becomes embedded:

- In operations
- In compliance
- In capital planning
- In occupant experience
- In long-term asset value

You are not selling a product.

You are asking someone to build their building's nervous system around you.

That level of dependency changes how decisions are made.

From Assumption to Expectation

For years, brand promises in this industry were implied:

"We've always been here."
"We're big enough."
"Our people know what they're doing."

That worked when change was slow.

It does not work anymore.

Today's market is shaped by:

- IT/OT convergence
- Cybersecurity pressure
- Cloud and edge architectures
- Data ownership concerns
- Sustainability mandates

Customers are no longer buying controls.

They are buying *future exposure.*

The Three Promises That Matter

Strong brands in building automation consistently deliver on three levels.

Performance — "It will work."

This is the baseline.

Reliability.
Uptime.
Security.
Scalability.

If this fails, nothing else matters.

Partnership — "We will stand with you."

Customers are not buying products.
They are buying relationships.

This shows up in:

- Support quality
- Training
- Documentation

- Channel stability
- Pricing transparency

The real question behind it:

"Will you still care after the sale?"

Progress — "We will not leave you behind."

Technology will change.
Standards will evolve.
Expectations will rise.

Customers want to know:

Will this still be relevant in ten years?

That means:

- Upgrade paths
- Backward compatibility
- Open architecture
- API strategy
- Ecosystem participation

If you cannot answer this, you lose strategic relevance.

What Happens When Promises Break

In this industry, failure doesn't disappear.

It spreads.

Through integrators.
Through consultants.
Through owners.
Through jobsite conversations.

People remember:

- Platforms that were abandoned
- Licensing models that changed suddenly
- Systems that claimed openness but weren't

Reputation compounds.

So does damage.

From Message to Discipline

A real brand promise cannot live in marketing.

It must show up in decisions.

If you claim openness:

- Your integrations must be real
- Your APIs must be usable
- Your contracts must support it

If you claim long-term partnership:

- You cannot casually abandon products
- You cannot hide roadmap changes
- You cannot treat partners as disposable

Every internal decision either reinforces or weakens the promise.

At Its Core, It's About Risk

Every buyer is managing risk:

- System failure
- Cyber exposure
- Vendor abandonment
- Integration breakdown
- Capital waste

A strong brand promise lowers perceived risk.

A weak one raises it.

When options look similar, the safer promise wins.

What Endures

In this industry, systems outlive leadership teams.

They outlive product cycles.

Sometimes they outlive the companies that installed them.

The brands that endure are the ones that leave behind:

- Systems that still run
- Architectures that still evolve
- Communities that still function

They keep their promises long after contracts expire.

Chapter 17 **Brand Loyalty and the Human Layer**

Brands do not live in logos.

They live in people.

In building automation, brand is carried by technicians, engineers, integrators, consultants, and operators—the people who make buildings work every day.

Where Brand Is Actually Built

Not in marketing meetings.

In mechanical rooms.
In job trailers.
In late-night troubleshooting calls.
In training sessions.
In moments when something isn't working and someone has to fix it.

That is where credibility is earned.

And that is where loyalty begins.

Trust Moves Through People

Customers do not trust platforms first.

They trust people.

The technician who solved the issue nobody else could.
The engineer who explained a complex integration clearly.
The support rep who stayed on the call until it was fixed.

That trust transfers.

From individual → to company → to platform.

That is how brand actually forms in this industry.

Brand Ambassadors Are Not Optional

Every person interacting with a customer becomes a brand ambassador—whether formally assigned or not.

And in a high-risk, long-lifecycle industry, those interactions carry weight.

Customers are asking:

Who stands behind this system?
Who will answer when something fails?
Who understands my environment?

They are not evaluating marketing.

They are evaluating people.

Loyalty Is Built Through Experience

Loyalty in building automation is not emotional in the consumer sense.

It is earned confidence.

It comes from repeated outcomes:

- Systems that work
- Integrations that hold
- Support that shows up
- Upgrades that don't break everything

Over time, that consistency becomes expectation.

And expectation becomes loyalty.

Community Is Not Marketing—It's Infrastructure

The strongest platforms build communities.

Not fan bases.
Working networks.

Places where:

- Problems are solved
- Knowledge is shared
- Best practices evolve

Over time, the community itself becomes part of the brand.

Because when someone says:
"If you're stuck, someone in the community has solved it..."

That is not marketing.

That is trust at scale.

Ecosystems Amplify or Destroy Brand

In this industry, no company operates alone.

Manufacturers.
Integrators.
Consultants.
Developers.
Owners.

Your brand is shaped not just by what you do—but by who represents you.

Strong ecosystems reinforce brand.

Weak ecosystems erode it.

The Role Has Changed

Today's brand ambassadors are not just technical implementers.

They operate at the intersection of:

- Engineering
- IT
- Business outcomes

Customers expect them to explain:

- Data strategy
- Cybersecurity implications
- Enterprise integration

The modern ambassador is not just installing systems.

They are translating complexity into confidence.

Culture Creates Ambassadors

You cannot script authenticity.

If the organization behind the product lacks discipline, clarity, or integrity—people in the field will expose it.

Strong culture produces strong ambassadors.

Weak culture produces inconsistent experiences.

And inconsistency kills brand.

What Actually Creates Loyalty

Loyalty is not blind.

Customers will evaluate alternatives.

They will test new technologies.

But when a platform continues to perform—when it continues to evolve without breaking trust—loyalty holds.

Because switching isn't just technical.

It's risk.

And people stay where they feel confident.

The Reality

In the end, brand in building automation is human.

It is:

- Who answers the phone
- Who shows up on-site
- Who takes ownership
- Who stays until it works

Technology enables the outcome.

People make it believable.

Chapter 18 **Brand as Strategic Capital**

Brand is often treated as decoration—something applied after the real work is done. A logo, a message, a campaign layered on top of products and operations. In infrastructure industries, that view is not just incomplete—it is wrong. Brand is not an accessory to the business. It is a strategic asset.

At its core, brand is accumulated trust. It reflects how consistently an organization performs across product delivery, support, partnerships, and decision-making over time. It is how the market remembers you. And in industries where systems operate for decades, memory carries weight.

This accumulated trust has real economic impact. Organizations with strong brands do not operate the same way as those without them. Sales cycles tend to be shorter because less time is spent overcoming doubt. Specification rates increase because engineers and consultants default to what they trust. Partners stay longer because relationships feel stable. Pricing pressure decreases because the conversation shifts from cost to confidence. Customers return because prior experience reduces perceived risk.

The underlying mechanism is simple: trust removes friction. When credibility exists, decisions move forward more easily. When it does not, every step requires justification. Technical capability still matters—but when multiple options appear similar, the trusted one wins.

This is why brand becomes more important as technology matures. Features can be replicated. Pricing can be matched. Specifications can be duplicated. What cannot be easily copied is confidence—the belief that a platform will continue to perform, evolve, and be supported over time. In building automation, that belief often outweighs incremental technical differences.

Brand also functions as a form of risk reduction. Every major decision in this industry carries long-term consequences. Buyers are not just evaluating what a system can do today—they are assessing what it might become over time. Will the platform still exist in ten or fifteen years? Will integrations remain intact? Will support continue? Will the ecosystem grow or shrink? These are not engineering questions alone. They are brand questions.

Over time, trusted platforms develop what might be called specification gravity. Consultants begin to specify them by default. Integrators prefer to work with them because they reduce complexity and risk. Owners request them because they have seen them perform across multiple projects. This momentum is not created through marketing campaigns. It is built through repeated, consistent outcomes.

That momentum creates leverage. Companies with strong brands do not start from zero in every conversation. Credibility precedes them. They enter new markets with less resistance. They face fewer objections. They are able to attract stronger partners and retain them longer. Brand, in this sense, compounds—it builds on itself over time.

This compounding effect extends beyond day-to-day operations into valuation. Investors evaluating technology platforms look for signals of durability: installed base loyalty, ecosystem strength, upgrade adoption, and market reputation. These are proxies for brand strength. They indicate whether a platform will continue to generate value in the future, not just in the present.

None of this happens by accident. Brand strength is a function of clarity and consistency. Strong organizations can answer fundamental questions without hesitation: why they exist, who they serve, what they stand for, and how they behave. That clarity aligns product decisions, partnerships, messaging, and long-term strategy. Without it, organizations fragment. With it, they reinforce themselves.

At a deeper level, brand is not separate from strategy—it is the visible result of it. It reflects how an organization makes decisions, how it treats customers and partners, and how it evolves over time. Every shortcut, every inconsistency, every broken promise leaves a trace. Over time, those traces accumulate into reputation.

In a technology landscape that is increasingly commoditizing, this distinction becomes critical. Hardware capabilities are converging. Connectivity is standardizing. Software features are becoming more uniform. As differentiation at the product level narrows, differentiation at the trust level expands. Brand becomes the deciding factor—not because it is more visible, but because it is more reliable.

The practical implication is straightforward. Organizations that build strong brands create optionality. They are less dependent on price competition. They can invest with a longer time horizon. They can shape markets instead of reacting to them. Organizations that neglect brand become interchangeable, and interchangeable companies compete primarily on cost.

Over time, the gap widens. One group builds compounding trust. The other chases short-term wins. One becomes easier to buy from. The other becomes easier to replace.

Brand, then, is not about perception alone. It is about performance sustained over time and recognized by the market. It is a form of capital—earned slowly, deployed continuously, and difficult to replicate.

In the end, the organizations that understand this do not treat brand as something to manage at the edges of the business. They treat it as something to protect at the center.

Chapter 19 **What I Learned: How a Framework Becomes a Brand**

There is a persistent belief in engineering-driven industries that brand does not matter.

The assumption is simple:
Engineers buy specifications.
Operators buy functionality.
Consultants buy performance.

That sounds rational.

It is also incomplete.

In infrastructure markets—especially building automation—brand matters more, not less. Because the cost of being wrong is not inconvenience. It is long-term consequence.

Where It Became Clear

The shift did not happen in a conference room.

It happened in the field.

As buildings became more connected and more digital, the questions changed. Owners stopped asking about points and controllers and started asking:

Can we see everything in one place?
Can we integrate systems that were never meant to connect?
Can we protect this investment long term?
What happens if we change service providers?

Those are not feature questions.

They are trust questions.

And trust is where brand begins.

The Gap the Industry Didn't See

At the time, the industry was not built around brand.

It was built around:

- Hardware specifications
- Protocol support
- Feature comparisons

Most companies were focused on products.

Very few were building identity.

But as building automation became infrastructure, something changed.

People needed something to trust—not just something that worked.

Watching Niagara Become Something Bigger

When Niagara emerged, it was not positioned as a brand.

It was positioned as a solution.

A way to connect systems that didn't talk.

But the deeper shift was architectural.

Instead of forcing replacement, it embraced what already existed.
Instead of competing with protocols, it spoke them.
Instead of owning the stack, it sat above it.

That decision mattered more than any feature.

Because it aligned with how buildings actually exist:

- Layered
- Fragmented
- Evolving over time

Niagara didn't try to simplify reality.

It respected it.

The Positioning That Changed Everything

Calling Niagara a *framework* was not semantics.

It was strategy.

Products get replaced.
Frameworks get built on.

That single distinction:

- Attracted OEMs
- Enabled integrators
- Reassured owners
- Allowed the ecosystem to grow

It moved Niagara above the churn of hardware cycles and protocol debates.

That is where durability begins.

What Brand Actually Became

Over time, one idea became clear:

Brand is accumulated behavior.

Not messaging.
Not campaigns.
Not positioning slides.

Behavior.

The support call that gets answered.
The upgrade path that still works years later.
The roadmap that shows up as promised.
The partner who is treated fairly.

In this industry, nothing is forgotten.

Every decision leaves a trace.

Over time, those traces become reputation.

And reputation becomes brand.

The Role of Leadership

Building a brand like this requires restraint.

Not just vision.

Restraint.

The discipline to:

- Not chase every trend
- Not overpromise capability
- Not sacrifice long-term trust for short-term revenue

I have seen platforms weaken because they moved too fast.

And I have seen platforms endure because they moved deliberately.

In infrastructure markets, speed impresses.

Consistency wins.

Stewardship Over Ownership

At some point, you realize something important:

A platform like Niagara is not owned in the traditional sense.

It is stewarded.

Because:

- Integrators build businesses on it
- Engineers design around it
- Owners depend on it

Breaking trust does not just impact a product.

It impacts an ecosystem.

That changes how decisions must be made.

Where This Is Going

The future will not make brand less important.

It will make it critical.

Because:

- Hardware is commoditizing
- Software capabilities are converging
- AI and analytics will be everywhere

When everything starts to look similar, something else determines who wins:

Belief.

Open vs closed.
Collaborative vs controlled.
Stable vs constantly reinvented.

The platforms that endure will not just be technically capable.

They will be philosophically consistent.

The Realization

In the end, brand is not decoration.

It is infrastructure.

And in a world where buildings are becoming digital systems, infrastructure is everything.

Chapter 20 **What Owners Actually Want**

For decades, the building automation industry has been optimized around systems.

Owners are optimized around outcomes.

That gap is where most of the frustration in this industry lives.

If you sit in enough meetings, attend enough conferences, or read enough product literature, you start to notice a pattern. The conversation revolves around platforms, protocols, dashboards, integrations, and features. The language is technical, and often impressive. But it is also disconnected from the way building owners and operators actually think about their world.

Because when you step into their position—even briefly—the priorities look very different.

Responsibility Changes Everything

From the owner's seat, the conversation is not about technology. It is about responsibility.

Owners and operators wake up every day asking a different set of questions:

- Are my buildings operating reliably?
- Are occupants comfortable and safe?
- Are we compliant with regulations?
- Are we secure—both physically and digitally?
- Are we controlling costs?

- What happens when something fails?
- And who is accountable when it does?

Technology shows up in those questions—but only as a means, never as the focus.

That distinction matters more than most of the industry acknowledges.

Because while vendors talk about features, owners think about risk.

They are not asking whether a system supports a specific protocol or API. They are asking whether it will keep their building running, reduce disruption, and make their operation more predictable.

Reliability. Simplicity. Accountability.

Those are not secondary concerns. They are the job.

Data Is Everywhere. Insight Is Not.

The industry often frames the future of buildings around data. And it is not wrong to do so.

Modern buildings generate enormous amounts of it. HVAC systems, lighting, metering, access control, environmental sensors, and cybersecurity tools all contribute to a growing stream of information. On paper, it looks like progress.

But from the owner's perspective, the problem is not a lack of data.

It is what happens next.

Data is scattered across systems.
Information is trapped inside vendor platforms.

Connections between systems are incomplete or fragile. And translating all of it into meaningful operational decisions is difficult at best.

So the real questions owners are asking are not technical ones. They are practical:

- Why is this building consuming more energy than the one next door?
- Which equipment is likely to fail next?
- Where should we invest capital to get the best return?
- How do we operate this portfolio more efficiently?

Those are not data questions. They are outcome questions.

And in too many cases, the industry stops one step short—delivering access to information, but not clarity.

The Problem No One Likes to Admit

There is another reality that becomes obvious when viewed from the owner's side.

They do not want more systems.

They already have too many.

A typical building or portfolio includes a patchwork of technologies:

- Building automation
- Energy management
- Metering
- Lighting control

- Security and access control
- Analytics platforms
- Cloud dashboards

Each one was introduced with a promise. Each one delivers some level of value. But each one also adds complexity—another interface, another contract, another integration point, another potential failure.

Individually, they make sense.

Collectively, they create friction.

From the outside, this looks like innovation. From the inside, it often feels like accumulation.

More to manage. More to maintain. More to reconcile.

And less clarity than expected.

The Noise

At the same time, owners are being asked to make decisions in an environment saturated with messaging.

Every company has an answer.

AI will transform your building.
The cloud is the only viable future.
IoT sensors will unlock everything.
Analytics platforms will solve your problems.
Digital twins will redefine operations.

Each message is compelling on its own. Each is supported by a plausible story.

But taken together, they create something else entirely.

Noise.

And when everything claims to be essential, it becomes difficult to determine what actually matters.

The result is not acceleration. It is hesitation.

Many owners arrive at a quiet conclusion:

If it is not clear what to believe, it may be safer not to act.

In that sense, the industry's effort to drive adoption can have the opposite effect—slowing it down through complexity and confusion.

We Have Been Solving the Wrong Problem

None of this is the result of bad intent.

The building automation industry has made enormous progress over the past several decades. Systems are more capable, more connected, and more powerful than ever before.

But along the way, the focus drifted.

We became very good at building systems.

We became less disciplined about what those systems were supposed to deliver.

Because buildings are not technology experiments.

They are assets.

They represent capital investment, operational risk, and long-term value. And when owners evaluate technology, they are not thinking about controllers, networks, or interfaces. They are thinking about outcomes:

- Operational reliability

- Cost control
- Energy performance
- Asset longevity
- Risk reduction
- Portfolio visibility

That is the scoreboard.

And when viewed through that lens, the disconnect becomes clear.

The Shift

If the goal is better outcomes, then adding more disconnected technology is not a solution.

It is part of the problem.

Which leads to a simple but uncomfortable realization:

The goal is not more technology—it's fewer systems delivering more outcomes. The Modern MSI exists to make that possible. Not by adding complexity, but by removing it.

That statement forces a shift in thinking.

From systems to outcomes.
From features to function.
From integration as an activity to integration as a means of simplification.

And it demands a different kind of role in the industry.

A Different Kind of Integrator

If owners want fewer systems, clearer insight, and better outcomes, then the traditional model of integration is not enough.

Connecting systems is necessary—but it is no longer sufficient.

The role has to expand.

Not into more technology, but into better orchestration.

A modern system integrator is not defined by the number of systems they connect, but by what those connections enable. Their responsibility extends beyond interoperability into structure, clarity, and usability of data. Beyond project delivery into ongoing operational value.

They are responsible for:

- Making data usable, not just accessible
- Aligning systems with operational goals
- Reducing complexity, not increasing it
- Enabling decisions, not just dashboards
- Supporting the building across its lifecycle, not just at commissioning

In simple terms, the role shifts from building a system to enabling a result.

From integration to outcomes.

What This Means for the Industry

This is not a small adjustment. It is a reorientation.

It changes how systems are designed.
How projects are delivered.

How success is measured.
And ultimately, how value is defined.

It also changes how companies position themselves.

Because brand, in this context, is no longer what is said in a presentation or printed in a brochure. It is what an owner experiences over time.

If the result of your solution is more complexity, more confusion, and more work to operate the building, that is your brand—regardless of how advanced the technology may be.

If the result is clarity, reliability, and measurable improvement in outcomes, that is your brand.

There is no separation between the two.

The Lens That Matters

At its core, this is not a technology discussion.

It is a perspective shift.

If we want to create real value in the built environment, we have to start with the people responsible for operating it. Not occasionally, and not as an afterthought—but as the primary lens through which decisions are made.

Because at the end of the day, the most important question is not:

What can this system do?

It is:

Does this make my building easier to operate, safer to manage, and more valuable over time?

That is the standard.

And it is the standard that will define what comes next.

Chapter 21 **When Infrastructure Becomes Identity**

The building automation industry did not begin with brand. It began with control—relays, controllers, wiring diagrams, and proprietary tools designed to keep buildings operating within defined conditions. Systems were installed, commissioned, and, if they performed as expected, they faded into the background. For a long time, that was sufficient. If the building worked, nothing else demanded attention.

As buildings became more complex, that model began to break down. Systems needed to communicate. Data needed to move. Owners wanted visibility across portfolios, not just within individual mechanical rooms. Integration shifted from an optional capability to a fundamental requirement. And once integration became central, architecture began to matter in a way it had not before.

What emerged from that shift was not just better technology, but a different category of importance. The systems connecting buildings became infrastructure. They were no longer tools used in isolation. They became the connective layer that allowed buildings to function as coordinated environments.

And then something less obvious—but more significant—began to happen.

Infrastructure became identity.

Not through marketing or positioning, but through repeated experience. When professionals encountered the same platform across projects, across vendors, and across portfolios, their relationship to it changed. It was no longer evaluated as a tool selected for a specific job. It became something assumed—a standard, a reference point, a baseline expectation. That is the moment when infrastructure crosses into brand.

The clearest signals of this shift did not come from strategy discussions or product roadmaps. They came from the field. Integrators, engineers, and operators began expressing the same underlying need in different ways. They wanted fewer environments to manage. They wanted systems that could work together without constant translation. They wanted architectures that allowed their teams to scale without relearning everything from project to project.

These were not requests for new features. They were expressions of friction—and a desire to remove it.

At the same time, another realization began to take shape. What was being built was not just a platform. It was a set of beliefs, reinforced through decisions over time. Openness instead of restriction. Community instead of control. Continuity instead of constant reinvention. Enablement instead of ownership.

Those beliefs were not declared as principles. They were demonstrated through behavior. And because they were consistent, they became recognizable.

That consistency is what allowed trust to form.

Niagara did not succeed by simplifying the industry. It succeeded by working within its complexity. Buildings are not uniform systems. They are layered environments, assembled over decades, containing multiple generations of technology from multiple vendors. Any platform that assumes a clean slate is misaligned with reality. Niagara aligned with reality instead. It created structure without forcing uniformity, and that alignment made it durable.

Over time, that durability produced something the industry does not easily manufacture: belief. Not belief in messaging, but belief in outcomes. That systems would integrate. That architectures would hold. That investments would not be stranded. That the platform would still be relevant years after the original decision was made.

In an engineering-driven industry, that kind of belief is not given lightly. It is earned through repetition.

That is why brand in this context is not superficial. It is the accumulation of decisions that hold up under pressure. It is the memory of what happened when things did not go as planned. It is the expectation that, when tested again, the outcome will be the same.

There is, however, a risk that comes with success.

As platforms gain momentum, the temptation is to accelerate—to expand faster, promise more, and pursue short-term opportunities that appear aligned with growth. In many industries, that approach is rewarded. In infrastructure markets, it is punished. Decisions made today persist for decades. Inconsistency introduced now compounds over time.

The brands that endure are not the ones that move the fastest. They are the ones that remain consistent under pressure. They resist unnecessary reinvention. They protect the expectations they have created. They understand that trust, once broken, is far more difficult to rebuild than it was to establish.

This is why brand cannot be separated from leadership. Every decision—product direction, pricing, partnerships, roadmap—either reinforces or weakens the expectations the market holds. Brand is not managed at the edges of the organization. It is shaped at the center.

Looking forward, the forces reshaping the built environment will only increase this dynamic. Artificial intelligence, advanced analytics, and data-driven operations will continue to expand what buildings are capable of. Hardware will become more interchangeable. Software capabilities will continue to converge. As technical differentiation narrows, trust will carry more weight, not less.

The platforms that endure in that environment will not be defined solely by what they can do. They will be defined by what they consistently represent.

Open or closed.
Collaborative or controlled.
Stable or constantly shifting.

Those distinctions will determine which platforms are trusted to carry buildings—and the organizations that depend on them—forward.

In the end, buildings will continue to do what they have always done. They will operate quietly in the background, supporting the people and organizations inside them. Mechanical rooms will still hum. Systems will still run. Most of the technology will remain unseen.

Some platforms, however, will not be invisible in the same way.

They will be trusted.

And the difference between those that are trusted and those that are replaced will not come down to features or specifications. It will come down to the accumulation of decisions over time—decisions that either reinforced confidence or eroded it.

That is what turns infrastructure into identity.

And identity, over time, into brand.

About the Author

Marc Petock

The Niagara Framework is regarded as foundational infrastructure in modern building automation, control and smart buildings. It is specified globally, taught in universities, embedded in product portfolios and trusted by owners, operators and specifiers worldwide. Through periods of growth and disruption, Niagara's identity and brand remained consistent: open, interoperable, vendor-neutral, and future-ready. This continuity reflected long-term brand governance rather than short-term marketing.

Marc Petock is widely recognized as one of the principal architects and marketers behind the Niagara Framework's global brand identity, ecosystem and positioning. During his leadership tenure as Vice President, Global Marketing at Tridium, Petock helped transform Niagara from an integration framework into the industry's dominant open platform and framework. His work focused on strategic positioning, community development, executive communications, and long-term brand governance.

He championed the concept of the "Niagara Community," embedding partner participation into the platform's value proposition. Under his stewardship, Niagara became not only a technical standard, but a professional and institutional one. Petock is known for his ability to translate complex system architecture into business relevant, outcomes, enabling enterprise-level adoption across portfolios and regions.

Today, Niagara functions as a professional standard and ecosystem identity in building automation and control. This level of institutional trust does not emerge spontaneously. It is cultivated through decades of disciplined stewardship.

Currently, Marc is Vice President, Chief Marketing & Communications Officer, and member of the Board of Directors at Lynxspring, where he leads corporate planning, and strategy for the company's product offerings, technology direction, defining and implementing marketing programs and initiatives and the budgets and people that support them.

In addition, he is responsible for creating the company's corporation and brand positions, their evolution and communications that support them.

Marc also serves as the spokesperson and industry thought leader where he often speaks at conferences and is a regular contributor of industry articles and papers. He is a founder of the Project Haystack Organization, an open standard for naming, tagging, and modeling data in building automation and smart building systems and member of the Board of Directors. He also serves as an Advisor to the Realcomm Organization.

.

Over the years, Marc has been recognized with several industry accolades including a Who's Who from M2M Magazine; a Niagara Community Spirit Award; the "Petock Award", an award named in his honor; Realcomm/iBCON Top 35 People to Watch, a Digital Impact Award (Digie) and the Realcomm/iBCON 2021 Lifetime Achievement Award.

Appendix **The Petock Smart Building Thesis**

Understanding the Transformation of the Built Environment

Every industry eventually reaches a moment where its underlying assumptions begin to change. What once seemed stable becomes transitional. What once seemed advanced became foundational. And what once seemed futuristic becomes inevitable.

The building automation and controls industry is now here.

For decades, building automation and controls focused primarily on controlling equipment. Systems were designed to regulate temperature, manage mechanical operations, and ensure basic operational reliability. Buildings were engineered systems first and digital environments second. That work was and remains essential. Reliable mechanical control is the foundation of building performance.

But something fundamental is changing. Buildings are no longer simply mechanical environments. They are becoming digital infrastructure. And digital infrastructure operates according to different rules. It evolves continuously. It generates data. It supports applications. It connects ecosystems of technology, services, and people. It requires security.

This transformation forms the foundation of what I describe as **the Petock Smart Building Thesis**—a framework for understanding how the built environment is evolving and where the industry is heading.

The Core Idea

At its simplest, the thesis can be summarized in a single idea: ***Buildings are transitioning from control systems to intelligent infrastructure platforms.*** This shift will reshape how buildings are designed, integrated, and operated.

Historically, building automation systems were built around three core objectives:

- Equipment control
- Operational efficiency
- System reliability

These goals remain essential. But they are no longer sufficient. Today's buildings must also support:

- Data-driven operations
- Portfolio-level intelligence
- Energy optimization
- Cybersecurity resilience
- Operational transparency
- Continuous system evolution

These capabilities require a fundamentally different architectural approach. They require platform thinking.

To understand this transformation, it helps to look at the four major perspectives that define the Smart Building Thesis.

The Four Forces Reshaping Building Automation

Four powerful forces are accelerating the transformation of the built environment.

1. The Rise of Data

Buildings generate enormous amounts of operational information—from temperature readings and equipment status to energy consumption patterns and environmental conditions.

Historically, much of this data remained trapped within control systems. Today it is increasingly being captured, structured, and analyzed to support more intelligent building operations. Operational data is becoming one of the most valuable assets within modern buildings.

2. Artificial Intelligence and Advanced Analytics

As building data becomes more accessible, advanced analytics and machine learning systems are beginning to uncover patterns that human operators cannot easily detect.

These technologies enable buildings to move beyond reactive control toward:

- Predictive maintenance
- Performance optimization
- Automated operational adjustments

Over time, artificial intelligence will enable buildings to operate with increasing autonomy.

3. IT, OT, AI Convergence

Building automation systems once operated largely separate from enterprise IT environments. Today that separation is disappearing. Modern buildings must integrate with:

- Enterprise data platforms
- Cybersecurity frameworks
- Cloud analytics systems
- Corporate operational dashboards

This convergence between Information Technology (IT), Operational Technology (OT) and Artificial Intelligence (AI) is fundamentally reshaping building system design.

4. Platform Ecosystems

Perhaps the most important shift is the emergence of technology ecosystems.

Instead of isolated vendor systems, modern building technologies increasingly operate within shared platform environments that allow multiple companies to innovate simultaneously.

Integration frameworks like the Niagara Framework helped introduce this model to the building automation industry. By enabling systems from multiple manufacturers to operate within a common architecture, Niagara created an environment where innovation could occur across an entire ecosystem rather than within a single vendor's product line.

This shift toward platform ecosystems is one of the defining characteristics of modern smart buildings.

The Evolution of Building Automation

The building automation industry can also be understood as progressing through several stages of technological maturity.

Each stage builds upon the capabilities of the one before it.

Stage One: The Control Era

The earliest generation of building automation focused on mechanical control.

Systems monitored sensors and operated equipment to maintain temperature, airflow, and comfort conditions.

These systems were typical:

- Proprietary
- Vendor-specific
- Isolated from other building systems

Buildings were controlled but not connected.

Stage Two: The Integration Era

The next major phase occurred as open protocols and integration frameworks began to emerge. Technologies such as BACnet, Modbus, LonWorks and Niagara allowed building systems to communicate with one another for the first time. Instead of isolated subsystems, buildings could now operate as integrated environments.

Stage Three: The Intelligence Era

Today the industry is entering what can best be described as the intelligence era. In this phase, the focus shifts from simply controlling and integrating systems to understanding and optimizing building performance through data. Key technologies driving this stage include:

- Edge computing
- Cloud analytics
- Artificial intelligence
- OT data platforms

Buildings are becoming data platforms.

Stage Four: The Autonomous Era

The next stage of building evolution is already beginning to emerge. In the autonomous building era, buildings will increasingly use artificial intelligence and advanced analytics to continuously adapt to changing conditions without constant human intervention.

Systems will:

- Predict operational issues
- Automatically adjust equipment
- Optimize energy performance
- Coordinate across entire building portfolios

The trajectory is clear. Control enabled integration. Integration enabled intelligence. Intelligence will enable autonomy.

Edge-to-Enterprise Architecture

As buildings evolve, their digital architecture must evolve as well. Modern buildings increasingly rely on a layered architecture that spans from physical equipment at the edge to enterprise-level intelligence platforms. This architecture typically includes several key layers.

Physical Infrastructure Layer

At the foundation are the physical systems that perform the mechanical functions of the building:

- HVAC equipment
- Lighting systems
- Meters
- Pumps
- Sensors

These systems generate the raw operational data of the building.

Edge Control Layer

Edge controllers manage real-time control of building equipment while collecting operational data. These devices ensure that equipment continues operating reliably while providing the data necessary for higher-level analysis.

Integration Layer

Integration platforms connect multiple systems together. They translate protocols, coordinate interactions, and create a unified operational environment across the building.

Frameworks like Niagara play a critical role at this level by enabling interoperability across vendors and technologies.

Data Management Layer

Once building systems are connected, operational data must be organized and structured. Independent data layers and OT data platforms normalize and prepare this information for analytics and enterprise use. This layer increasingly defines how building data becomes accessible across organizations.

Enterprise Intelligence Layer

At the highest level of the architecture sit enterprise applications. These systems use analytics, machine learning, and visualization tools to provide insights that guide operational decisions across entire building portfolios. This is where building data becomes actionable intelligence.

The Smart Building Value Stack

As buildings become more digital, the sources of value within building automation are shifting upward. Historically, most value resided in the physical equipment and control systems that operated it.

Mechanical reliability and performance were the primary goals. But as buildings become more connected and data-driven, value is increasingly created at higher layers of the technology stack.

Equipment and Control

Mechanical performance remains essential. But it is now only the starting point.

Integration Platforms

Integration frameworks create value by allowing multiple systems to coordinate and share information. Without integration, building data remains fragmented.

Data Infrastructure

Data platforms create value by organizing operational information and making it accessible across systems and organizations. This enables new levels of visibility into building performance.

Analytics and Intelligence

At the highest level, analytics and artificial intelligence generate insights that help operators:

- Improve energy performance
- Extend equipment life
- Reduce operational costs
- Optimize building performance

The industry is moving from mechanical optimization to informational optimization. Buildings are no longer simply machines that provide comfort. They are becoming data-driven environments that continuously adapt to improve operational outcomes.

The Larger Implication

Taken together, these perspectives reveal a fundamental shift within the built environment. Buildings are evolving from static infrastructure into dynamic platforms. They will increasingly operate as interconnected systems that generate and utilize data to improve performance, efficiency, resilience, and sustainability.

Technology platforms, integration frameworks, and data architectures will play an increasingly central role in enabling this transformation. For system integrators, technology providers, engineers, and building owners alike, understanding this shift is essential.

The built environment is entering a new era. An era where buildings are no longer passive assets.

They are intelligent, connected, and continuously evolving platforms. And the organizations that understand this transformation will help define the next generation of smart buildings.

www.ingramcontent.com/pod-product-compliance
Ingram Content Group UK Ltd.
Pitfield, Milton Keynes, MK11 3LW, UK
UKHW062311290726
14090UKWH00018B/1007